1992 Yearbook Supplement to McGraw-Hill's NATIONAL ELECTRICAL CODE® HANDBOOK

ASSISTANT EDITORS

Robert A. Germinsky

Managing Editor
edi (Electrical Design and Installation) Magazine

Brendan A. McPartland

Facilities Engineer
Coldwell Banker

Steven P. McPartland

Sales Engineer
Northern Electric Supply

Jack E. Pullizzi

Electrical Specialist
AT&T Bell Labs

CONTRIBUTING EDITORS

Rex Cauldwell

Owner
Little Mountain Electric & Plumbing

Robert W. Salter

Applications Engineer
Bussman Div./Cooper Industries

Steve Schaffer

Applications Engineer
Bussman Div./Cooper Industries

Charles E. Small, PE

Owner
Atlantic Instrumentation

1992 Yearbook Supplement to McGraw-Hill's NATIONAL ELECTRICAL CODE® HANDBOOK

Joseph F. McPartland

Editorial Director/Publisher
edi (Electrical Design and Installation) *Magazine*
452 Hudson Terrace
Englewood Cliffs, NJ 07632

Brian J. McPartland

Associate Editor/Publisher
edi (Electrical Design and Installation) *Magazine*
452 Hudson Terrace
Englewood Cliffs, NJ 07632

McGraw-Hill, Inc.

New York St. Louis San Francisco Auckland Bogotá
Caracas Lisbon London Madrid Mexico Milan
Montreal New Delhi Paris San Juan São Paulo
Singapore Sydney Tokyo Toronto

2 3 4 5 6 7 8 9 0   DOC/DOC   9 8 7 6 5 4 3 2

ISBN   0-07-045914-2   {HC}
ISBN   0-07-045913-4   {PBK}

The sponsoring editor for this book was Harold B. Crawford, the editing supervisor was Alfred Bernardi, and the production supervisor was Suzanne W. Babeuf. It was set in Century Schoolbook by McGraw-Hill's Professional Book Group composition unit.

Printed and bound by R. R. Donnelley & Sons Company.

NATIONAL ELECTRICAL CODE® *is a registered trademark of National Fire Protection Association, Inc., Quincy, Massachusetts, for a triennial electrical copyrighted publication of such corporation. The term* NATIONAL ELECTRICAL CODE *as used herein means the publication constituting the* NATIONAL ELECTRICAL CODE *and is used with the permission of National Fire Protection Association, Inc. This yearbook supplement does not emanate from and is not sponsored nor authorized by the National Fire Protection Association, Inc.*

Contents

An Invitation from the Authors

The 1992 Yearbook Supplement to McGRAW-HILL's NATIONAL ELECTRICAL CODE® HANDBOOK was developed to help you resolve the many Code-related issues that electrical professionals face on a daily basis. You the reader bring to this *Yearbook Supplement* valuable personal experiences in working with the Code that can be shared with your colleagues. We therefore invite you to describe in letter form ways in which you have resolved problems in applying Code requirements. We also invite you to ask questions regarding Code interpretation that have been troublesome. Such questions will be answered in future *Yearbook Supplements*.

Letters are to include your full name, address, and affiliation, which will be withheld on request should your letter be reprinted in the *Yearbook Supplement*. Such letters become the property of McGraw-Hill.

ABOUT THE AUTHORS

JOSEPH F. McPARTLAND was the Editorial Director of *Electrical Construction and Maintenance, Electrical Construction and Maintenance Products Yearbook, Electrical Wholesaling*, and *Electrical Marketing Newsletter*. He is now Editorial Director/Publisher of *edi (Electrical Design and Installation)* Magazine. He has authored twenty-six books on electrical design, electrical construction methods, electrical equipment, and the *National Electrical Code®*. For over 42 years he has been traveling throughout the United States conducting seminars and courses on the many aspects of electrical design, engineering, and construction technology for electrical contractors, consulting engineers, plant electrical people, and electrical inspectors. Mr. McPartland received a Bachelor of Science degree in Electrical Engineering from Thayer School of Engineering, Dartmouth.

BRIAN J. McPARTLAND was an associate editor of *Electrical Construction and Maintenance* and has more than 16 years' experience in electrical technology. He is now Associate Editor/Publisher of *edi (Electrical Design and Installation)* Magazine. After serving in the U.S. Navy Submarine Force, during which time he earned his degree in electrical technology, Brian held positions in both product engineering and sales with various electrical equipment manufacturers. He is coeditor of McGraw-Hill's *Handbook of Practical Electrical Design* and coauthor of *McGraw-Hill's National Electrical Code® Handbook* and EC&M's *Illustrated Changes in the 1990 National Electrical Code®*.

1992 Yearbook Supplement to McGraw-Hill's NATIONAL ELECTRICAL CODE® HANDBOOK

Compliance with the NEC Only Provides the Minimum

Sec. 90-1(b). Sec. 90-1(b) essentially indicates that compliance with NEC rules is the bare minimum. But, simply satisfying NEC rules may still leave the designer and/or installer legally exposed.

We are all aware that there has been a tremendous increase in the amount of litigation leveled against electrical designers and installers. Generally, these legal actions are civil suits charging negligence. However, in one unfortunate incident, the defendant, an electrical contractor, was indicted and convicted of criminal manslaughter. As a result of this upward trend in electrical litigation, many electrical professionals are looking for ways to protect themselves from the potentially devastating consequences of being found guilty in the event a civil or criminal charge is filed against them—or even from the stigma and negative publicity of being charged.

One would think that complying with prevailing codes and standards—such as the National Electrical Code (NEC), Underwriters Laboratories (UL), American National Standards Institute (ANSI), Institute of Electrical and Electronic Engineers (IEEE), etc.—would certainly provide protection against a successful prosecution. But this is not always true. For example, as most are aware, the NEC is a *minimum* standard. If the NEC has the force of law (i.e., if it has been adopted by a federal, state, county, or municipal legislative body, which it *is*, virtually everywhere across the nation), compliance with the requirements of the NEC represents only the barest minimum level of safety legally required. In some cases, as electrical professionals, it is necessary for each of us to determine if a Code rule is really adequate with respect to accomplishing the desired safety benefit in any specific instance and what else, if anything, should be done to assure the intended safety benefit. Let's look at an example.

In a number of its rules, the NEC requires a warning or caution sign to further provide that an electrical installation is "essentially free from hazards." In a number of instances, the wording required by the Code for field-installed warning or caution signs does not include an instruction that indicates what action must be taken by the person(s) reading the sign. Although the NEC would not require such additional information, it has been repeatedly ruled in court that failure to provide such information is the same as not providing the sign itself. In fact, in one particular million-dollar suit, the jury found in favor of the plaintiff because the plaintiff's lawyer convinced them that a drop-light should have had a marking prohibiting its use in an auto-repair shop.

In addition to assuring compliance with the "legal interpretation" regarding warning and caution signs, going beyond that which is minimally required by the NEC is known as a "good faith" gesture. That is, because the installer made an additional effort to assure safety and performed in "good faith," this action will normally be taken into consideration by the court and can serve to demonstrate that the installer made every conceivable effort to meet the *intent* of the rule as well as the *letter* of the rule. The Occupational Safety and Health Administration (OSHA) places great emphasis on "good faith" safety efforts. In a number of instances, "good faith" actions on the part of the installer were viewed favorably and resulted in lower fines.

Regardless of the type of prosecution, in virtually every one of the cases that I know of or have been involved with, the plaintiff (or prosecuting) attorney requests a trial by jury. The jury will invariably be made of people who are not capable of fully and clearly understanding the technical issues involved. As a result, whenever the defendant can show that additional steps, no matter how seemingly insignificant, were voluntarily taken—that "good faith" was exercised in the design and/or installation—this action will generally be viewed in very favorable light by the jury. And the more such instances that can be demonstrated, the better. Indeed, it may be the only evidence that the jury really *can* evaluate.

With the foregoing in mind, it should be clear that in the never-ending battle to limit legal exposure and liability, acting in "good faith" can be the difference between an acquittal or guilty verdict should you ever become the defendant in a legal action.

Inasmuch as any additional efforts will benefit the owners, the costs for these "extras" should be passed along to them. If the benefits are pointed out to the owners, in most cases they will be willing to pay. If not, think of the additional outlay as inexpensive insurance against the potential horrors associated with a lawsuit. To better protect yourself and your business use your best judgment and, to the maximum extent possible, employ the "good faith" advantage.

Code Enforcement: Specific Rules Should Be Referenced!

Sec. 90-4. The wording of Sec. 90-4 limits the inspector's authority to *interpreting* other Code rules.

In Sec. 90-4 of the NEC, the Code makes clear that responsibility for interpretation of the rules and regulations given within the NEC rests with the "authority having jurisdiction"—generally, the local electrical inspector. And that section is widely interpreted to mean that the local electrical inspector has the final say regarding what is and what is not acceptable.

From a practical standpoint, there is a very definite need to designate an ultimate authority capable of rendering an immediate decision during an electrical inspection. If no one were so designated, disagreements between designers/installers and inspectors regarding interpretation of NEC rules could go on forever. And, because the only means of obtaining a "Formal Interpretation" on an NEC rule from the National Fire Protection Association (NFPA) takes a considerable amount of time (sometimes months), use of that approach for resolving any disagreement is economically prohibitive. Can you imagine putting a project on hold for months while waiting for an answer? Obviously not! Therefore, to provide for an immediate resolution to any disagreement regarding interpretation, the rule of Sec. 90-4 establishes the local electrical inspector (the "authority having jurisdiction") as the final judge of what is required.

During our constant travel and contact with electrical designers and installers throughout the country, we hear of horror stories where an inspector required this or that, and when a question was raised by the designer and/or installer, Sec. 90-4 was cited as the reference. In fact, it seems as if everyone involved in electrical design and installation has either firsthand or word-of-mouth knowledge about some application where an inspector refused to pass the installation until such-and-such was done "the way I want it." Although designers and installers acknowledge the need for, and respect the rule of, Sec. 90-4, when an inspector cites that section as the basis for "doing it my way," many feel that the inspector has overstepped the authority granted by Sec. 90-4.

But is that true? Does Sec. 90-4 place limits on the inspector's authority? From the wording used it would seem so.

The key portion of Sec. 90-4 that needs to be examined is the second sentence which states: "The authority having jurisdiction will have responsibility for making interpretations of the rules." That wording would certainly seem to imply—if not flatly dictate—that the inspector's authority is limited to *interpreting* specific Code rules. Therefore, an inspector needs to tell the designer/installer what rule has been violated according to his or her interpretation. Remember, the concept for application and enforcement of the NEC revolves around the prohibitions and requirements spelled out by specific Code rules. Basically stated, a designer or installer can do anything, unless it is specifically prohibited or they are required to do otherwise. Therefore, if the application in question satisfies all NEC requirements and does not violate any prohibition, then it should be considered acceptable, even if the application seems unusual or unconventional. Just because an application is unconventional, it does not mean it is unsafe or contrary to Code.

Fortunately, many electrical inspectors already interpret Sec. 90-4 in this manner and they do provide specific reference(s) for any violation(s) cited, and, in some cases, they even make recommendations about what might be done to bring the installation into compliance with the referenced Code section(s). Even if recommendations are not offered, where a specific rule is referenced, it gives the designer and/or installer something concrete upon which to make a decision regarding what might be the most economical means available to achieve compliance. In fact, in many cases, the designer and/or installer may not have been aware of the rule. And in those cases, the inspector serves to help limit the legal exposure of the designer and/or installer while assuring that the installed electrical system is essentially safe.

Obviously, such an approach will require an intimate knowledge of the NEC on the part of the inspector. And such knowledge can only be obtained by constant study and analysis of the NEC and actual field installations. But, this is no less than what is expected of the designer and installer.

Given the concerns for safety and economic consequences of holding up a job, inspectors today, more than ever, must be extra sharp and thoroughly familiar with all Code rules. When one considers the awesome responsibility and the very critical function of the electrical inspector in terms of assuring safe installations, it is easier to understand why an inspector may revert to the use of Sec. 90-4 as a reference. Generally, this occurs when the inspector is not exactly sure about the acceptability of a specific application. However, Sec. 90-4 itself can not be used as the basis for "doing it my way." Whenever a violation is issued, a specific rule should be referenced.

Use of "Listed" Equipment Is Virtually Mandatory

Secs. 110-2 and 110-3. One of the most common questions regarding application of distribution and utilization equipment in electrical systems revolves around whether or not such equipment is *required* to be "listed," that is, evaluated by a nationally recognized testing laboratory in accordance with a recognized standard. Although for certain applications the NEC does specifically require the use of "listed" equipment, in many other rules words such as "suitable" or "identified" are used to define which products or materials are acceptable for a given application. Although there is no doubt that listed equipment must be used when it is specifically called for by a Code rule, what about when equipment or conductors are only required so as to be "suitable" or "identified"?

As far as the NEC is concerned, in general, use of "listed" equipment is not mandatory. Indeed in NEC Sec. 110-2, it is only required that any equipment or conductors used be "approved." The definition of that word, as given in NEC Article 100, indicates that equipment or conductors are "approved" if they are acceptable to the authority having jurisdiction—the local electrical inspector. That is, if the local inspector has evaluated a product or material for use in a given application and is satisfied that it possesses adequate mechanical and electrical properties to perform safely under the specific conditions of use, the local inspector may permit an unlisted component to be used. However, as covered in Sec. 110-3(a)(1), listing or labeling by a third-party testing lab may also be used as the basis for acceptance. As a result many inspecting authorities require that equipment be listed before they will "approve" its use. And that means if the local inspector *requires* listed products—which is the inspector's prerogative—then so does the NEC. But, what if the inspector does *not* require the use of only listed equipment? Is it permissible to use other than listed equipment?

One can completely satisfy the requirements of the NEC using nonlisted equipment if the equipment or conductors are acceptable to the local inspector and no other rule specifically requires listing. But if the installation falls under the jurisdiction of the Occupational Safety and Health Administration (OSHA), compliance with its rules and regulations must be assured.

The OSHA regulations regarding the installation and maintenance of electrical systems and equipment are covered in Subpart S of 29 CFR. Although the sequence of references is somewhat convoluted, it is the intent of that standard to generally *require* all equipment and conductors to be "accepted, certified, listed or labeled by a nationally recognized testing laboratory." (See the definition of "Acceptable" as given in section 1910.399(a)(i) of CFR 29, Subpart S.) There are, however, certain exceptions to that basic requirement.

One exception is granted for equipment and conductors of a type that no nationally recognized testing lab recognizes. That is, if a particular piece of equipment is a type of equipment that no third-party testing lab lists or labels, then use of a listed piece of equipment is not mandatory. But, where such a piece of equipment is used, it must be evaluated by someone. In addition, it seems as if some documentation of the evaluation procedure must be produced and retained by the building or facility owner. In a similar manner, when a custom-made piece of equipment is used, it is not required to be listed or labeled, but again, some testing or safety examination must be conducted and documented.

What does all this mean? Simply that use of equipment listed for the specific application is *always* acceptable, whereas use of nonlisted equipment may or may not be. And, if the OSHA rules apply to the installation—which they do to all places of employment, whether in commercial, industrial, institutional, or, in some cases, residential applications—then nonlisted equipment must be of a type that no third-party testing lab lists or labels, or it must be a custom piece of equipment. And, in either of these two cases, some safety evaluation must be performed by the inspecting authority or another federal, state, or municipal agency.

As we now see, the NEC strongly recommends the use of listed equipment and conductors. And OSHA, as well as many local electrical inspectors, insist on the use of listed equipment. Given the continually expanding requirements for use of listed equipment by the NEC and the mandatory OSHA or local inspection authority's requirement that, to the maximum extent possible, only listed equipment be used, the reality for electrical designers and installers today is that use of listed equipment is virtually mandatory.

Rating of Fuse and Fuseholder

Secs. 110-3(b), 240-3, Ex. No. 4.   Here's a summary of the correct application data for the rating of fuses and fuseholders.
A 200A, 600VAC fusible switch was chosen as the disconnect for a 208VAC feeder to supply a *continuous* nonmotor load of 185A. First of all, Sec. 220-10(b) of the NEC requires that the rating of the overcurrent device protecting a feeder supplying a continuous load must be "not less than...125 percent of the continuous load." For a 185A continuous load, the minimum rating of feeder protection had to be 1.25 × 185A, or 232A. Based on that, the minimum size of standard fuse that would have to be used is a 250A fuse (the next higher standard rating of protective device above 232A). Design analysis proceeded as follows:
A continuous load of 185A should require use of a 200A switch. In the UL "white book" (General Information Directory), under "Enclosed Switches," the UL recognizes that "Switches without fuseholders (unfused) have been tested to determine their acceptability for continuous operation at their marked rated load." That says that a 200A nonfusible switch can be properly used to handle a "continuous" (operating steady for over 3 hours) load of up to 200A.
But the very next sentence in the "white book" says that "Fused enclosed switches are marked 'Continuous load current not to exceed 80 percent of the rating of fuses employed in other than motor circuits.'" Very clearly, that rule of UL would require the 185A continuous nonmotor load being served here to be no more than 80 percent of the rating of the fuses in the switch. Using 250A, 250VAC fuses in the switch (they fit into the fuseholders intended for 200A, 600VAC fuses) satisfies that rule—because 80 percent of 250A is 200A, well above the 185A load. Note that this UL requirement is the same as—but the reciprocal of—the rule of NEC Sec. 220-10(b), as discussed above.
Now here's where a question arises: May the minimum required rating of 250A, 250VAC fuses be used in a 200A, 600VAC switch? If the load is not going to exceed 185A, a 200A switch would seem to be suitable—especially because a nonfusible switch assembly itself is *rated* for a 200A continuous load and a fusible switch assembly is tested (using copper dummies) for temperature rise with a 200A continuous load. And the actual 185A continuous load does *not* exceed 80 percent of the 250A fuses used in the switch.

Or, must the 250A fuses be used in a 400A switch? That would entail vastly higher material and labor costs than the use of a 200A switch.

In the past, manufacturers have shown in their specification and application literature that a 100A switch may be used with 125A fuses when applied with a 100A continuous load (so that the 100A load is not in excess of 80 percent of 125A fuses). *But this issue is no longer addressed.*

The accompanying photograph shows the application described here, as it was designed and installed at a facility of a prestige industrial. The installation has operated effectively and safely for 17 years. The conductors used are No. 4/0 THW copper, with an ampacity of 230A—which are properly protected by the 250A fuses, as the next

250A fuses in a 200A switch feeding a 185A continuous load —is this acceptable?

Because a 400A switch costs twice that of a 200A switch, and because a decision could involve a large number of feeder switches on a given job, the very large additional costs involved in using a 400A switch instead of a 200A switch dictate very careful analysis. The concern for economics is important, *but* compliance with applicable codes and standards must be given highest priority. And this type of analysis and decision making must be brought to the task of selecting any equipment for modern electrical systems.

higher standard rating of protective device above the conductor's 230A ampacity NEC Sec. 240-3, Ex. No. 4. (Or, 4/0 THHN copper, with an ampacity of 260A, could be used.)

In the view of the designer, this application is not a violation of any NEC, UL, or NEMA rules. All of us are (or ought to be) committed to full and detailed compliance with *all* industry safety standards. But that compliance often must be provided within the tightest bounds of economy.

(All of the calculations and equipment sizings given in this example would apply to any continuous load from 181A up to 200A.)

The cost comparison is important because if the continuous load can be reduced to 180A, it could be handled by a 225 CB—which is a totally different consideration from a 400A fusible switch in terms of both material and labor costs. In fact, if a 400A switch has to be used for any fuse rated over 200A, then any continuous load above 161A would require a protective device rated at 125 percent of its value, and therefore would require a 400A switch.

There is a specific violation in the application discussed here. Can you spot it? Or do you see additional objections? [*Comment*: This entire matter was put out for response in a major trade publication. Selected (edited) responses follow.]

In response to the question about suitability of the fusible switch application as shown, it was noted that it was a violation of NEC Sec. 240-60(b), which says:

"Fuseholders shall be so designed that it will be difficult to put a fuse of any given class into a fuseholder that is designed for a current lower, or voltage higher, than that of the class to which the fuse belongs."

Violation of Sec. 240-60(b) was noted by Leo Delaney, Marshfield, Massachusetts, and Eric David, I.C.B.O. Certified Electrical Inspector, Long Beach, California.

Some people argue that the wording of Sec. 240-60(b) is not direct and clear. First, the electrical systems designer or the installing electrician—who are responsible for satisfying the code rule—do *not design* fuseholders. To tell them "Fuseholders shall be so *designed...*" seems to be a misdirected requirement. And then, the way the rule is worded, the Code *would be satisfied* if it is simply "difficult to put" the 250A, 250V fuses in the 200A, 600V fuseholders. Because of the thicker blades on the 250A, 250V fuse, it *is* "difficult" to put it in the 200A, 600V fuseholder, which is designed for a thinner fuse blade. But that ought *not* be the basis for Code acceptability of a particular application.

A far more direct and understandable basis for the unacceptability of the 250A, 250V fuses in the 200A, 600V fuseholders is given by the UL regulations that apply. In the UL "Green Book" (the Electrical Construction Materials Directory) and in the UL "White Book"—under the heading "FUSES—CARTRIDGE NONRENEWABLE"—it says "Nonrenewable fuses of a given voltage rating or current rating range are not interchangeable in the same fuseholders with fuses of a different voltage or current rating range" (page 52 in the 1990 UL White Book).

The UL regulation was the basis for violation cited by Alan Nadon, Certified Electrical Inspector, Electrical Department, City of Elkhart, Indiana. He would also cite the fused switch under discussion as violating NEC Sec. 110-3(a) and (b)—especially 110-3(a)(7). He stated, "Just because it physically fits does not make it right. The improper contact area between fuseholder and fuse blade may result in heat and a failure. There are circuit breakers that fit in panels of another manufacturer but are not listed for such use and should not be accepted any more than a piece of copper tubing used in place of a fuse."

Another reader who based his objection on the UL rule was Glenn Zieseniss, Crown Point, Indiana, who noted that "the blades are not fully engaged in the fuse clips." He also stated, "I was recently on a River Boat (not covered by the NEC) out of Davenport, Iowa, and noted the same type of installation for the air-conditioning units on the top deck."

Paul Duks of Underwriters Laboratories, Northbrook, Illinois, forwarded a UL response to the application of the 250A, 250V fuses in the fusible switch:

Another question was asked about how to choose the proper fusible switch size: "May the minimum required rating of 250 A, 250 VAC fuse be used in a 200 A, 600 VAC switch?" The answer is no.

The article stated that 250 A, 250 VAC fuses will fit into the fuseholders intended for 200 A, 600 VAC fuses. This is only possible if the fuse is improperly forced into the clips. Doing so is likely to cause damage to the fuseholder, poor contact at the blades, and high termination temperatures. A 200 A, 600 V fuseholder is designed for 0.188-inch thick blade, while the 250 A, 250 V fuse's blade is 0.25-inch thick. This excess thickness will bend apart the fuse clips. Also, the 200 A, 600 V fuseholder is made to receive a 9–5/8-inch long fuse while the 250 A, 250 V fuse is 8–5/8-inches long. The fuse will not fit properly; if one blade is fully in the clip, the other blade will have only about one-fourth the contact area it should.

Section 240-60(b) of the NEC requires that fuseholders be so designed that it will be difficult to put a fuse of any given class into a fuseholder that is designed for a current lower or voltage higher than that of the class to which the fuse belongs. The objective of this rule is to assure fuses are used in circuits for which they are rated. Fuses and fuseholders have been designed so that a 250 A, 250 V fuse will fit properly only into a 400 A, 250 V fuseholder. In all installations, the fuse and fuseholder must be of the *same voltage rating marked on the fuseholder* and proper ampere range.

Do You Know All About "Better Protection" for Motor Starters?

Sec. 110-10. UL-listed Type E starters and Type 2 protection better satisfy the rules of Sec. 110-10.

All over the nation, as commercial-industrial electrification continues to grow at an ever-accelerating pace, electrical designers and installers are finding motor control to be one of the most controversial areas of technology. The ever wider use of new, larger, and more expensive motors focuses clearly on the need for better control of the motor loads and *better protection* of the very heavy financial investments those motors represent.

It has always been necessary to evaluate a motor starter for its ability to start and stop a motor, to do so without serious voltage dip or instability, to provide control of motor speed, if needed, and—one of the most critical tasks, so commonly performed by overload relays in the starters—to provide effective protection against damage to the costly motor. But another concern of rapidly growing importance is the need to assure that the short-circuit protective device used for the motor branch circuit will either minimize or even completely prevent any damage to the controller itself if a short-circuit or ground-fault occurs in the motor circuit on the load side of the starter.

We all know that damage to a starter is caused by energy—generated either thermally (heat) or magnetically (force). By far, the most important damage is thermal, which can weld contacts closed and damage the running overload relays. That thermal energy is commonly expressed as I^2t—indicating that the heat is proportional to the square of the fault current that flows and to the time duration from initiation of the fault to opening and clearing of the current flow by the protective device. It is strictly the speed of operation of the protective device that can prevent damage, by holding t to a very small value, and in the case of current-limiting devices, by also reducing the RMS-value (the heating effect) of the I. We must constantly focus on that basic reality.

In recent years, high visibility has been given to those considerations in use of combination motor starters—a single, unitized assembly that incorporates: (1) motor circuit safety disconnect (usually a switch or CB); (2) motor short-circuit and ground-fault protection (usually fuses in the switch or the CB); (3) motor controller (usually a contactor); and (4) running overload protection (usually overload relays connected into the contactor control circuit). One or more (even all) of those functions can be combined into a single device—like a fused switch or CB. But combinations of those functions are now made available in some totally new designs of combination starters by manufacturers of so-called IEC (International Electrotechnical Commission) starters—the smaller, more compact, and often more sophisticated starters popularized through effective marketing in the United States by European control manufacturers. And it is these new designs that must be carefully studied by all designers and installers, to become more knowledgeable about the technological concepts behind these units and the terminology that is applied to them.

Do you know what an "E-type" starter is? What about a "self-protected starter?" Could you give a clear, simple explanation of the difference between "Type 1 protection" and "Type 2 protection"? Among even the best engineers and electricians, we have found only a very hazy and doubtful understanding.

In the conventional combination motor starters we have all been using for years, it has been acceptable to the NEC and UL if slight damage is done to the starter contacts and heater coils by let-through energy when the short-circuit protective device (fuses or CB) clears a short or ground fault on the load side of the starter. The problem is pinpointed in NEC Sec. 110-10, where a short-circuit protective device *only* has to protect against "excessive damage" to controller components. Damage less than "excessive" (whatever that is) is permitted by the NEC and the UL testing. According to IEC standards, that is "Type 1" protection.

IEC standards has another designation, "Type 2" protection—which applies where a combination starter has such fast-acting short-circuit protection that *no* damage is done to the starter components on a fault. Starters of that type listed by the UL are referred to as "self-protected" starters; and under UL standard 508, they are referred to as having "Type E" construction—from which comes the phrase "E type."

During the past year, some important new concepts have been introduced to the everyday tasks of selection and application of effective control of motor loads. One concept involves a radically new type of combination motor starter, known as the "Type E *starter*." The Type E starter is also called a "self-protected combination motor starter" by certain manufacturers and is referred to as a "self-protected control device" by Underwriters Laboratories (UL). Basically stated, self-protected motor starters combine all of the NEC-required protection and control capabilities for a motor branch circuit within a single device. The other concept involves selection of short-circuit protection ahead of the motor starter to prevent significant damage to the starter under short-circuit conditions, and that protective coordination is called "Type 2 *protection*."

There have been numerous references—by both trade journals and manufacturers—to Type E starters and Type 2 protection. But, very few designers and installers actually know what those designations mean. Indeed, many believe those terms to be interchangeable. Although the two terms are related, they have different origins and meanings.

Type E starters

Combination motor starters submitted by manufacturers for listing by Underwriters Laboratories (UL) are evaluated in accordance with the provisions of UL Standard 508. And that standard recognizes a number of different configurations for a combination motor starter. That is, different devices may be used to provide a similar function. For example, the short-circuit and ground-fault protection may take the form of a fuse in a fusible switch, a conventional inverse-time CB, a fusible device known as a motor short-circuit protector (or MSCP), or an instaneous-trip CB. Depending on the actual components used, each starter will be designated by a UL letter-code that corresponds to its "Construction Type." That is, the specific means provided to achieve the desired protection and control functions will be indicated by the UL "Construction Type" letter-codes for combination motor starters (see Figure 1).

There are five letter-code designations that indicate the "construction type" of UL-listed starters. The most commonly used combination starters are the fusible-switch combination motor starter (Type A) and the inverse-time CB combination motor starter (Type C). The fusible MSCP, which is not widely applied, is designated as Type B, and the MCP, which is a listed combination starter with an instaneous-trip CB, is referred to as Type D. Those letter designations covered all UL-listed combination motor starters, that is, up until recently. The newly-listed self-protected starters are designated as Type E construction, hence the term "Type E" motor starter.

As indicated earlier, the Type E (self-protected) starter is a device that combines all the NEC-required motor-circuit protective and control functions in a single unit. That is, a Type E starter will contain the necessary components to provide: (1) a motor disconnecting means; (2) motor branch-circuit short-circuit and ground-fault protection; (3) motor overload protection; and (4) ON/OFF control of the motor (Figure 2). Although starters of UL-designated Construction Types A through D also perform all those functions, they are not performed

Type "E" starters are a new category of combination motor starters that combine short-circuit and ground-fault protection, overload protection, disconnecting means, and a contactor within a single, integrated device. Today's commercially available listed Type E or "self-protected" starters provide a level of coordination and protection that is very similar to what the IEC refers to as "Type 2 protection" in IEC Standard 947.4. The net result is "better" protection of the *starter,* which is more consistent with the wording of **NEC** Sec. 110-10.

by a single, integrated device. But that is not the only difference between Type E starters and other "types."

Because of their physical configuration, today's commercially available Type E starters are evaluated using a more stringent set of pass/fail criteria for the short-circuit test than other construction-types of UL-listed combination motor starters. For all "other" types of

TABLE 79.2
VARIOUS CONSTRUCTIONS OF
COMBINATION MOTOR CONTROLLERS

Component Parts[a]	Construction Type				
	A	B	C	D	E[b]
Disconnecting Means	Manual Disconnect (UL 98 or UL 1087)	Manual Disconnect (UL 98 or UL 1087)	Circuit Breaker (UL 489)	Circuit Breaker (UL 489)	Self-protected Control Device (UL 508)
Short-Circuit Protective Device	Fuse (UL 198)	Motor Short-Circuit Protector (UL 508)	Inverse Time Trip Circuit Breaker (UL 489)	Instantaneous Trip Circuit Breaker (UL 489)	
Motor Controller	Magnetic (UL 508)	Magnetic (UL 508)	Magnetic (UL 508)	Magnetic (UL 508)	
Overload Protection	Overload Relay (UL 508)	Overload Relay (UL 508)	Overload Relay (UL 508)	Overload Relay (UL 508)	

a Tests are conducted on the individual components per the applicable UL standards shown in the parentheses in the table following each component. The UL standards are as follows:

UL Number	Designation
98	Enclosed and Dead-Front Switches
198 series	Fuses
489	Molded Case Circuit Breakers
1087	Molded Case Switches

Table 79.2 revised September 27, 1989

b See paragraph 79.2.

Figure 1 A reproduction of Table 79.2 from UL Standard 508 (15th edition, adopted 6 June 91). As can be seen, each "type" of combination motor controller is assigned a letter-designation that indicates the specific components used to perform the four NEC-required control and protective functions. The newest combination starter is the self-protected or "Type E" combination starter. The design of listed Type E starters commercially available today is such that a load-side short-circuit within the device's interrupting rating will not damage the starter's contacts or overload device.

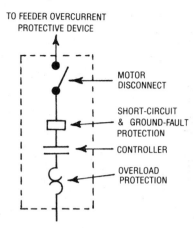

TO FEEDER OVERCURRENT
PROTECTIVE DEVICE

MOTOR
DISCONNECT

SHORT-CIRCUIT
& GROUND-FAULT
PROTECTION

CONTROLLER

OVERLOAD
PROTECTION

Figure 2 One-line diagram of the NEC-required control and protective functions provided by a Type E combination motor starter. In addition to simplifying selection of motor circuit protection and control, today's commercially available Type E starters provide a greater level of short-circuit coordination, which eliminates damage to the starter's contacts and overload device.

combination starters, welding of the ON/OFF control contacts is permitted. And for Types B, C, and D combination starters, the replaceable overload element (overload heater) is permitted to be destroyed when subjected to the UL short-circuit test. Although the actual wording of the short-circuit test pass/fail criteria for "self-protected control devices" would permit similar damage to a Type E starter, because today's commercially available Type E (self-protected) motor starters use the same contacts for starting/stopping and either disconnection or branch-circuit protection, the contacts must *not* sustain such damage. For today's Type E starter, after sustaining the maximum short-circuit for which the device is rated, the contacts must be capable of operating. The wording of the UL pass/fail criteria would also permit "burnout of the current element" (i.e., overload heater). However, the listed Type E starters now on the market use bi-metal overload devices and such overload devices must *not* be damaged or need recalibration. That is, due to their design, the Type E starters now listed are *required* to be operable, without any loss of calibration to the overload protective device, after the starter has been subjected to the short-circuit test. Starters of the other construction types may be damaged to the point where the contactor and/or overload device(s) would have to be replaced (Figure 3).

Some may be thinking, "You mean all of the UL-listed combination motor controllers I've used over the years could be damaged to the point of control contacts welding under a fault condition within the combination motor controller's short-circuit interrupting rating?" The answer is yes. The next question is, "How does that wash with the rule of Sec. 110-10?" That one is not as easy to answer.

In NEC Sec. 110-10, the Code requires that the capabilities and lim-

58.27 TEST CONDUCTED WITH FUSES — After the protective device has cleared the fault, an overload relay or industrial control equipment incorporating an overload relay shall comply with the following:

A. The contacts shall not disintegrate — evaporate — but welding of contacts of the controller is considered acceptable. Damage to other parts of the controller that would impair their function shall not occur.

B. Burnout of a pigtail lead, ignition of the cotton, or any other risk of a fire shall not occur.

C. The overload-relay base, temperature sensing element, or other parts shall not be damaged.

Exception: Damage to such parts may occur if the current elements burnout in accordance with the exception to item E.

D. The fuse connected between the live pole and the enclosure shall not open.

E. The current element shall not burn out or be damaged to the extent that it would not perform in accordance with the calibration requirements in paragraph 56.2.

Exception: A current element for use with a motor rated less than 0.25 amperes may burn out when subjected to the test described in paragraph 58.1 employing a 1-ampere fuse.

F. The door or cover shall not be blown open and it shall be possible to open the door or cover. Deformation of the enclosure is acceptable, but shall not result in the accessibility of live parts as determined by the use of the rods specified in paragraph 5.28.

Paragraph 58.27 revised October 19, 1990

58.28 TESTS PERFORMED WITH INVERSE-TIME OR INSTANTANEOUS-TRIP CIRCUIT BREAKERS, OR COMBINATION MOTOR CONTROLLER SELF-PROTECTING CONTROL DEVICE — After the protective device has cleared the fault, an overload relay or industrial control equipment incorporating an overload relay shall comply with the following:

A. The contacts shall not disintegrate — evaporate — but welding of the load switching contacts of the controller is considered acceptable. Damage to other parts of the controller that would impair their function shall not occur. Contacts that serve as disconnection or branch circuit protection means shall not weld.

Exception No. 1: Damage to the overload relay base or other parts may occur if current element burnout occurs.

Exception No. 2: A combination motor controller self-protected control device module may contain load switching means that have welded provided that the module is located on the load side of the disconnection and branch circuit contacts and that the module can be replaced.

B. Burnout of the current element is acceptable.

C. The fuse connected between the live pole and the enclosure shall not be open.

Exception: The fuse may be open if current element burnout occurs. However, additional tests are to be conducted (1) with the current element that burns out but with a solid electrical connection — without fuse — between the enclosure and live pole, and (2) with the current element jumpered or with a current element sufficiently large that burnout does not occur.

D. The door or cover shall not be blown open and it shall be possible to open the door or cover. Deformation of the enclosure is acceptable, but shall not result in the accessibility of live parts as determined by the use of the rods specified in paragraph 5.28.

Paragraph 58.28 revised October 19, 1990

Figure 3 Reproduction of the short-circuit test pass/fail criteria for industrial control devices protected by fuses (left) and those protected by other overcurrent devices from UL 508. As can be seen in paragraph A of both categories, contact welding as a result of the short-circuit test *is* acceptable. And, for other than fuses, the overload heaters are permitted to be destroyed (paragraph B). Although the Exception No. 2 under paragraph A for other than fuse protection (at right) would appear to permit the contacts of a self-protected starter to weld, today's listed Type E starters use the same contacts for ON/OFF control *and* either disconnect or branch-circuit protection. Therefore, the last sentence in paragraph A prohibits the welding of their contacts. And because those listed devices also employ bi-metal overload devices, without heater elements, no damage is permitted to the overload device.

itations of all NEC-required overcurrent protective devices be evaluated and coordinated to assure fault clearing within its rating "without extensive damage to the electrical components of the circuit." Isn't welding of the contacts and destruction of the overload device considered to be "extensive damage"? Generally, one would be inclined to take that view. But such damage is permitted for *listed* combination motor starters, because it is *not* considered to be "extensive." Therefore the rule of Sec. 110-10 is satisfied.

Type 2 protection

As we have seen, the Type E motor starter has special construction requirements and is capable of operating and protecting the motor against overload after sustaining a short circuit within its rating. Those requirements are spelled out in Part E of UL Standard 508 for self-protected control devices.

The UL requirement that self-protected motor starters be capable of operating, essentially, as if they were new after a short circuit, is very similar to the protective characteristics that are required by the International Electrotechnical Commission Standard IEC 947 and known as "Type 2" protection. That is, completely independent of the UL, and independent of UL designations and terminology, the Inter-

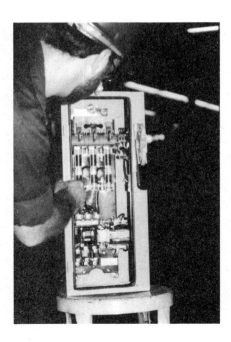

For fusible switch type combination motor starters, Type 2 protection (sometimes called Type 2 coordination) can be achieved by selecting fuses that have been found by *testing* to have time-current opening characteristics that limit let-through energy to a level that will *not* damage any of the starter's components—including its contacts or overload device—when the combination starter sustains a short-circuit within its listed interrupting rating. Such Type 2 protection can be achieved for both NEMA starters and the so-called IEC starters, *but* the specific type and rating of fuse to be used must be that designated by the manufacturer of the starter. Each manufacturer can submit their starter(s) for the necessary testing to obtain UL verification of the coordination for each starter and fuse combination that meets the standards set for Type 2 protection.

national Electrotechnical Commission describes a nearly identical protection criteria and refers to it as "Type 2 protection." Basically stated, when a device provides Type 2 protection, the let-through energy is so low that downstream (load-side) components will not be damaged by the fault current they're exposed to. And there are commercially available devices and combination of devices that meet the IEC Type 2 protection standard.

What should be understood is that today's listed self-protected (Type E) motor starters will essentially provide IEC Type 2 protection. That is, all listed Type E combination motor starters provide a level of protection that can be classified as the equivalent of IEC Type 2 protection. If self-protected starters are now available and provide a level of protection more consistent with the wording of Sec. 110-10, does that mean that only self-protected starters should be used to assure compliance with the NEC requirement to prevent "extensive damage"? The answer to that is no, not really.

In addition to the development of the self-protected starter, a number of motor control manufacturers have been testing combination starters of the other UL-listed construction types to determine the correct rating of overcurrent device that when used in combination with their starter, would provide the equivalent of IEC Type 2 protection. That is, where the manufacturer-recommended overcurrent protection is provided on the line-side of the starter, there will be no damage to the starter's contacts or overload device due to a short circuit within the motor controller's short-circuit interrupting rating.

As far as future motor control applications are concerned, it would seem that use of either a listed Type E combination starter or a listed combination starter of another construction type with the manufacturer-recommended line-side (upstream) overcurrent protection that provides Type 2 protection should be the standard approach. Although it is true that IEC Type 2 protection is *not* mentioned or specifically required anywhere in the NEC, the wording of Sec. 110-10 can be taken to require the "equivalent" of IEC Type 2 protection. If one were to use one of the other listed type combination starters without additional protection to prevent damage to the contacts or overload device, one could certainly be open to criticism. That is, if an accident occurs, you may be asked to explain why you didn't take steps to prevent what others interpret to be "extensive damage" to the motor starter. While in the past such a question generally would not have been raised, now that the technology to prevent *any* damage is commercially available, failure to apply this new technology and protection concept will be very difficult to explain.

In addition to the concerns for limiting legal exposure, each of us should consider the benefits of providing "better protection" for the

"Type 2" motor starter protection—Is it required by the **NEC?** Literally, there is *no* **Code** rule requiring that the short-circuit protective device(s) in a combination motor starter must protect the starter contacts and overload relays against *any* disabling damage due to short-circuit or ground-fault current—which is the idea behind the concept known as "Type 2 protection," or as it is sometimes called, "Type 2 coordination." In fact no **Code** rule even refers to "Type 2" protection. **NEC** Sec. 110-10 simply requires protection against "*extensive*" damage" to the components of a starter. Because there is no definition of "extensive damage" given in the **Code,** electrical inspectors have to provide their own definition. Two inspectors have told us they would consider "extensive" to be any level of damage that is disabling, that is, any damage greater than the damage permitted by Type 2 protection. Such an interpretation would make Type 2 application a mandatory **Code** requirement. Be sure to follow this controversy in the Technical Committee Report (TCR) discussion about the rewording of Sec. 110-10 for the 1993 **NEC.**

starter. That idea may seem somewhat foreign to many because the motor is generally the only component that most are concerned with protecting. However, the self-protected and other Type 2-protected starters can provide greater reliability and operating continuity, which will help to reduce downtime. Because all electrical designers and installers should constantly be striving for improved reliability and operating continuity of electrical systems, use of self-protected (UL Type E) starters, and Type 2-protection for all other starters, should be standard practice in all future motor control applications.

Dedicated Space ABOVE Switchboards and Panelboards

Secs. 110-16(a), 384-1, and 384-4. Area above switchboards and panelboards must be free of "foreign piping."

Question: I have an area that *is not* an electrical room. It's what we call the "incinerator room." It's big, clean, well lighted, etc. Within this room, I relocated electrical equipment to an area that meets all of the clear workspace requirements given in Sec. 110-16(a).

After I had relocated the electrical equipment, I found out that a 6-in. gas main was to be installed directly over the electrical equipment. The bottom of this gas main was also less than 25 ft. from the floor.

I stopped the job and explained the situation to my boss. Then I showed him Sec. 384-4 where it states that switchboards and panelboards must be installed in "rooms or spaces" dedicated to the equipment and in addition to the clear workspace described in Sec. 110-16(a) "such space...shall include an exclusively dedicated space extending for 25 ft. from the floor to the structural ceiling." I also pointed out that this rule prohibits installation of any "piping, ducts, or equipment" that is "foreign" to the electrical equipment within this dedicated *space.*

My boss is of the opinion that the rules of Sec. 384-4 *only* apply to equipment installed in *electrical rooms.* I say no, these rules apply to switchboards and panelboards whether installed in an electrical *room* or not. Can you help?

Answer: From the way you described the installation in your question, I would say that the rules of Sec. 384-4 *do* apply and that the installation of the gas piping within the dedicated space above any switchboards and panelboards is a literal violation of the wording used in this Code section.

The wording of this rule and its four exceptions has created much confusion among electrical people as to its intent and correct application in everyday electrical work.

The original intent of this rule was to eliminate the then common practice of installing water piping—chilled-water pipes, steam pipes, potable water pipes, sanitary clean outs, and other such pipes that carry liquids—above switchboards and panelboards and thereby prevent the destructive arcing and violent burndown that can be caused by water or other fluids leaking from the piping down onto the equipment. However, the actual wording of this rule excludes "piping, ducts, equipment foreign to the electrical system or architectural appurtenances" from being installed in, entering, or passing through the dedicated space.

Inasmuch as the literal wording of this rule does not differentiate between water- or liquid-carrying piping and piping for air or gases, and the fact that the wording also excludes foreign equipment and architectural appurtenances from the dedicated space, the installation of *any* "foreign" piping above equipment covered by Article 384 is prohibited. And the first sentence of Sec. 384-1 clearly indicates that this Article covers "switchboards, panelboards, and distribution boards...for light and power...and battery charging panels." (Remember, although the initial intent of the rule may have been more limited in scope, the final wording of the rule covers additional situations and Code rules must be interpreted and enforced as written.) Individual switches, CBs, motor control centers, and all other electrical equipment are *not* subject to the rules of Sec. 384-4.

The designated equipment covered by the rule (switchboards, panelboards, etc.) does *not* have to be installed in a dedicated room, although it may be (See FPN No. 1 to Sec. 384-4). As an alternative to a dedicated room, the rule of Sec. 384-4 also finds it acceptable if the equipment is installed in dedicated "spaces." *But* the "space" must be exclusively dedicated to the equipment.

Color-Coding of Conductors
Provides Enhanced Safety

Secs. 210-4 and 210-5. **Color-coding of conductors provides enhanced safety and satisfies the requirement for "identification" where more than one voltage system is used in buildings, on structures, etc.**

Branch-circuit conductors are regulated by NE Code rules on color-coding or other identification of specific conductors of the circuit. Such rules are given in Sec. 210-5 and Sec. 210-4. Code rules on color-coding of conductors (Sec. 210-5) apply only to branch-circuit conductors and do not directly require color-coding of feeder conductors. However, NEC Sec. 384-3(f) does require identification of the different phase legs of feeders to panelboards, switchboards, etc.—and that requires some technique for marking the phase legs. Many design engineers do insist on color-coding feeder conductors to provide ready and sure identification that will enable effective balancing of loads on the different phase legs. Some require full color-coding for the entire length of conductor insulation, but others will accept colored-tape banding at all terminations of black-insulated conductors.

Color-coding of branch-circuit conductors can be divided into three categories.

1. Grounded conductor (usually a "neutral" conductor). The grounded conductor of a branch circuit (the neutral of a wye system or a grounded phase of a delta) must be identified by a continuous *white* or *natural-gray* color, for the entire length of conductors No. 6 or smaller.

Where wires of different systems (such as 208/120 and 480/277V) are installed in the same raceway, box, or other enclosure, the neutral or grounded wire of one system must be white or natural-gray (usually the 208/120V system neutral); and the neutral of the other system must be white with a thin color line (other than green) running along the white insulation. For two systems of different voltage installed in the same raceway or enclosure, the wording of the rule of Sec. 210-5(a) is best interpreted to require the neutral of one system to be colored either white or gray and the neutral of the other system to be colored white with a color line (other than green) along the white insulation. The code rule wording might be taken to prohibit use of one white neutral and one gray neutral for two different voltage systems installed in the same raceway or enclosure. If there are three or more systems in the same raceway or enclosure, the additional neutrals must be white with different-colored tracers (other than green). The point is that neutrals of different systems must be distinguished from each other when they are in the same enclosure.

Exceptions to Sec. 200-6 modify the basic rule that requires use of continuous white or natural-gray color along the entire length of any insulated grounded conductor (such as a grounded neutral) in sizes No. 6 or smaller. Exception No. 3 permits use of a conductor of other colors (black, purple, yellow, etc.) for a grounded conductor in a *multiconductor cable* under certain conditions:

a. That such a conductor is used only where qualified persons supervise and do service or maintenance on the cable—such as in industrial and mining applications.

b. That every grounded conductor of color other than white or gray will be effectively and permanently identified at all terminations by distinctive white markings or other effective means applied at the time of installation.

This permission for such use of grounded conductors in multiconductor cable allows the practice in commercial and industrial facilities where multiconductor cables are commonly used.

2. Hot conductors. Sec. 210-5 contains no rule requiring that individual hot (ungrounded) conductors of a multiwire circuit be identified. It is no longer necessary, for instance, to use black, red, and blue for phase conductors of a circuit, with a white conductor as a neutral. Absence of a specific code rule on required color for phase (or hot) legs of branch circuits permits all phase conductors (hot legs) of a circuit to be the same color. That is, all circuit conductors could be black, or blue, or red, or any other color. Or the conductors may have any combinations of different-color insulations. Such application, from the standpoint of effective design, makes branch circuits difficult to install and load—as well as potentially unsafe.

3. Equipment grounding conductor. An equipment grounding conductor of a branch circuit (if one is used) must be color-coded green or green with one or more yellow stripes—or the conductor may be bare, as covered in Sec. 210-5(b). However, Exception No. 1 refers to Sec. 250-57(b), which permits an equipment grounding conductor larger than No. 6 to be other than a green-insulated conductor or a green-with-yellow-stripe conductor.

Sec. 250-57 permits an equipment grounding conductor with insulation that is black, blue, or any other color—provided that one of the three techniques specified in Sec. 250-57 is used to identify the conductor as an equipment grounding conductor. The first technique consists of stripping the insulation from the insulated conductor for the entire length of the conductor appearing within a junction box, panel enclosure, switch enclosure, or any other enclosure. With the insulation stripped from the conductor, the conductor then appears as a bare conductor, which is recognized by the code for the purpose. A second technique that is acceptable is to paint the exposed insulation green for its entire length within the enclosure. If, say, a black insulated conductor is used in a conduit coming into a panelboard, the length of the black conductor in the panelboard can be painted green to identify it as an equipment grounding conductor. The third acceptable method is to mark the exposed insulation with green-colored tape or green-colored adhesive labels. Green-colored conductors must not be used for any purpose other than equipment grounding.

More than one voltage level

Sec. 210-5 of the NEC—entitled "Color Code for Branch Circuits"— contains no mandatory requirements for color-coding of the ungrounded conductors (the "hot" legs, or "phase" legs) of a branch circuit. But absence of a requirement for color-coding is significant *only* for a "building" that has an electrical system operating at only *one* voltage level—208Y/120V or 240/120V (Figure 1). But if a build-

WHEN BUILDING CONTAINS ONLY
ONE SYSTEM VOLTAGE FOR CIRCUITS:

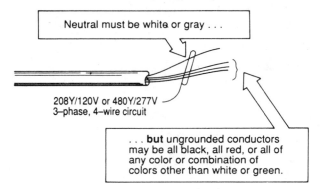

Neutral must be white or gray . . .

208Y/120V or 480Y/277V
3–phase, 4–wire circuit

. . . **but** ungrounded conductors may be all black, all red, or all of any color or combination of colors other than white or green.

Figure 1 When a building contains an electrical system that operates at only *one* voltage level for its service, feeders, subfeeders, and branch circuits, the code simply requires that the grounded conductor (usually a neutral) be identified—and there is no code rule requiring identification of the ungrounded conductors (the hot legs, or phase legs). However, identification of those ungrounded conductors is a "must" for effective electrical design, *and* such identification should always be by color-coding of the entire length of conductors.

ing has circuits operating at two or more voltage levels, the rules of Sec. 210-4(d) must be satisfied. In that Section, the rule for multiwire branch circuits requires some means of identification of each of the hot (ungrounded) conductors of branch circuits in a "building" that contains wiring systems operating at two or more different voltage levels (Figure 2). Note that the rule applies *only* to "conductors of branch circuits that are installed" in a "building." Literally, this requirement would not apply to outdoor electrical systems that utilize two or more voltage levels.

This rule makes it necessary to identify phase legs of circuits when more than one voltage system is used for multiwire branch circuits in a building. For instance, a building that utilizes both 208Y/120V circuits and 480Y/277V circuits must have separate and distinct color-coding of the hot legs of the two voltage systems—*or* must have some means other than color-coding, such as tagging, marking tape (color or

IF THERE ARE TWO SYSTEM VOLTAGES:

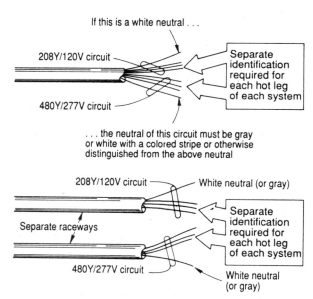

Figure 2 Separate identification of ungrounded conductors is required only if a building utilizes more than one nominal voltage system. Neutrals must be color-distinguished if circuits of two voltage systems are used in the same raceway [Sec. 210-5], but not if different voltage systems are run in separate raceways.

numbers), or some other identification that will satisfy the inspecting agency. And this rule further states that the "means of identification must be permanently posted at each branch-circuit panelboard"—to tell how the individual phases in each of the different voltage systems are identified (Figure 3).

The wording of the rule requires that the "means of identification" must distinguish between all conductors "by phase and by system." But, if a building uses only one voltage system—such as 208Y/120V or 240/120V single-phase—then no identification is required (although it should be used) for the circuit phase (the "hot" or ungrounded) legs. And where a building utilizes two or more voltage systems, the separate, individual identification of ungrounded conductors must be done whether the circuits of the different voltages are run in the same or separate raceways (Figure 4).

```
BRANCH CIRCUITS ARE IDENTIFIED BY COLOR CODING.

BLACK---A PHASE                    BLUE----C PHASE

RED-----B PHASE                    WHITE---NEUTRAL
```

```
BRANCH CIRCUITS ARE IDENTIFIED BY COLOR CODING.

BROWN----A PHASE                   YELLOW---C PHASE

ORANGE---B PHASE                   GRAY-----NEUTRAL
```

```
BRANCH CIRCUITS ARE IDENTIFIED BY COLOR CODING.

BROWN----A PHASE                   YELLOW---C PHASE

ORANGE---B PHASE                   WHITE w/tracer-----NEUTRAL
```

Figure 3 If a building's electrical system has circuits operating at two or more voltage levels, "each branch-circuit panelboard"—and that means "every" branch circuit panelboard at the two or more voltages—must be provided with some type of sign or label (such as one of the above) that tells the "means" of identification of the ungrounded conductors. Such a sign should say something like, "Phase conductors are distinguished by color-coding: BLACK—phase A; RED—phase B; BLUE—phase C." In each case, the sign must explain the specific differentiation of conductors according to both "phase" and "system." The above are three suggested labels that could cover the vast majority of cases.

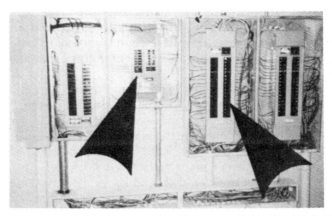

Figure 4   Sec. 210-4(d) requires that "the means of identification shall be permanently posted at each branch-circuit panelboard" of any building electrical system that operates at two or more voltage levels. But the rule does not indicate where on a panel the identification must be posted. Although a sign label (such as shown in Figure 3) would be readily evident on the inside of the panel door, where it would be readily inspected, the label might be put on the interior of the panel enclosure, where it would be obvious to people working on the circuits and would be protected from unintentional removal.

Color-coding—a safety "must"

Color-coding of circuit conductors (or some other method of identifying them) should be a mandatory design requirement for all ungrounded conductors (phase legs of *all* circuits) as well as for grounded conductors (such as neutrals) and for equipment grounding conductors. This important matter deserves the close, careful, complete attention of all electrical people.

By providing ready identification of the two or three phase legs and neutrals in wiring systems, color-coding is the easiest and surest way of differentiating the conductors from each other to provide for balancing loads among the phase legs, thereby providing full, safe, effective use of total circuit capacities.

In circuits where color-coding is missing or not effectively applied, loads or phases get unbalanced, many conductors are either badly underloaded or excessively loaded, and breakers or fuses are often increased in size to eliminate tripping due to overload on only one phase leg. For reasons of safety and energy conservation as well as full, economic application of system equipment and materials, modern, safe, electrical systems demand the many real benefits that color-coding can provide.

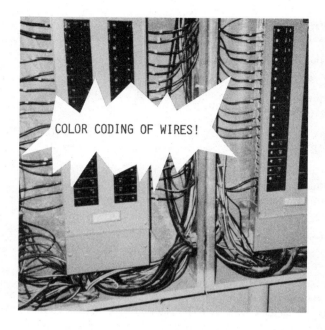

COLOR CODING OF WIRES!

Over the greater period of its existence, the NEC required a very clear, rigid color-coding of branch circuits for good and obvious safety reasons. Color-coding of hot legs to provide load balancing is a safety matter. Sec. 220-4(d) requires balancing of loads from branch-circuit hot legs to neutral. Sec. 220-22 bases the sizing of feeder neutrals on clear knowledge of load balance to arrive at the current value that constitutes "maximum unbalance." And mandatory differentiation of voltage levels is in the safety interests of electricians and others maintaining or working on electrical circuits, to warn of different levels of hazard.

Because 99 percent of electrical systems involve no more than two voltage configurations for circuits up to 600V, and because there has been greater standardization in circuit voltage levels, there should be industry-wide standardization on circuit conductor identifications. A clear, simple set of rules can be established as design requirements for the preponderant majority of installations, with exceptions made for the relatively small number of cases where unusual conditions exist and a special color-coding is demanded. Color-coding should follow some basic pattern, such as the following:

120V, 2-wire circuit: grounded neutral—white; ungrounded leg—black.

240/120V, 3-wire, single-phase circuit: grounded neutral—white; one hot leg—black; the other hot leg—red.

208Y/120V, 3-phase, 4-wire: grounded neutral—white; one hot leg—black; one hot leg—red; one hot leg—blue.

240V delta, 3-phase, 3-wire: one hot leg—black; one hot leg—red; one hot leg—blue.

240/120V, 3-phase, 4-wire, high-leg delta: grounded neutral—white; high leg (208V to neutral)—orange; one hot leg—black; one hot leg—red.

480Y/277V, 3-phase, 4-wire: grounded neutral—white with a color tracer; one hot leg—brown; one hot leg—orange; one hot leg—yellow.

480V delta, 3-phase, 3-wire: one hot leg—brown; one hot leg—orange; one hot leg—yellow.

By making color-coding a set of simple, specific color designations, standardization will assure all of the safety and operating advantages of color-coding to all electrical systems. Particularly today, with all electrical systems being subjected to an unprecedented amount of alterations and additions because of continuing development and expansion in electrical usage, conductor identification is a regular safety need over the entire life of the system.

A second step in clarification and expansion of color-coding would be a mandatory design requirement in electrical specifications and plans for color-coding of *all* circuits—multiwire branch circuits, branch circuits without a neutral or other grounded conductor, feeders, and even service conductors. As indicated in Sec. 384-3(f) of the NEC for "Phase Arrangement" of busbars in switchboards and panelboards, correct, effective loading of all circuits—to get full capacity of all phases and to prevent unknown balances with the attendant chance of oversizing of protection and overloading of conductors—depends upon ready phase identification of all conductors at all points in a system.

Of course, there are alternatives to "color" identification throughout the length of conductors. Color differentiation is worthless for colorblind electricians. For them, the word *black, red, blue, white, brown, orange, yellow,* or *gray* can be repeatedly printed along the conductor insulation to positively identify each hot leg or neutral of 208/120V and 480/277V systems. And it can be argued that color identification of conductors poses problems because electrical work is commonly done in darkened areas where color perception is reduced even for those with good eyesight. The NEC already recognizes white tape or paint over the conductor insulation-end at terminals to identify neutrals where conductors are larger than No. 6 (Sec. 200-6). Number-markings spaced along the length of a conductor on the insulation (1, 2, 3, etc.)—particularly, say, white numerals on black insulation—have proved very effective for identifying and differentiating conductors. Or the letters A, B, and C could be used to designate specific phases. Or a combination of color and marking could be used. But some kind of conductor identification is a design essential to assure safe, effective hookup of the ever-expanding array of conductors used today.

Countertop Receptacles

Secs. 210-8(a)(5) and 210-52(c). **Receptacles in fixtures installed in a kitchen must be supplied from the "small appliance" branch circuits, and may require GFCI protection.**

Question: I have a few related questions about the installation of a surface-mounted fluorescent fixture that contains an integral 125V, 15A receptacle above a countertop under a cabinet in the kitchen of a dwelling unit.

If the fixture receptacle is supplied from one of the two small appliance branch circuits and provided with GFCI protection where the receptacle is within 6 ft. of a kitchen sink, can this receptacle serve one of the "required receptacles" described in Sec. 210-52(c)? If not, do I still have to supply the fixture receptacle from one of the required two small appliance branch circuits and/or provide GFCI protection for the fixture receptacle if it's within 6 ft. of a kitchen sink?

Answer: There are actually two questions here. I will answer them in the order asked.

The answer to the first question regarding acceptance of a receptacle that is an integral part of a lighting fixture as a "required receptacle" is no. This is covered by the last paragraph of Sec. 210-52(a) where it says: *The receptacle outlets required by this section shall be in addition to any receptacle that is part of any lighting fixture or appliance, located within cabinets or cupboards, or located over 5½ ft. above the floor.*

This rule was adopted in the 1975 edition of the NEC to clarify that the Code does not recognize receptacles that are part of a lighting fixture or appliance as satisfying the rules for "required receptacles" in dwelling units. The substantiation for this change pointed out that such receptacles should not be recognized because if the lighting fixture or appliance is replaced, it may be replaced with another unit that does not have an integral receptacle. This may encourage or necessitate the use of extension cords, which is essentially what these rules are intended to discourage or eliminate.

Although the rule given in this last paragraph of Sec. 210-52(a) may appear to apply only to those receptacles required by part (a) of Sec. 210-52, this requirement is generally interpreted to apply to *all* receptacles required by *all* subparts of Sec. 210-52.

When this statement was included in the 1975 NEC, it was added in what was then Sec. 210-25(b). At that time, Sec. 210-25(b) covered all

receptacles required in "Dwelling Type Occupancies." Then, in the 1981 edition, these rules were renumbered and relocated to new Part C of Article 210. The reason given in the first sentence of the substantiation submitted with the proposal for this revision stated: *For the purpose of clarification, consolidate the basic rule in one thought and all exceptions in another.* And the rules of Secs. 210-52(a) through (f) were adopted for the 1981 edition of the NEC to achieve this goal.

The substantiation indicates it was the intent of the revised rules to clearly require compliance with all provisions of the "basic rule" [Sec. 210-52(a)] unless modified by one of the "exceptions" [Secs. 210-52(b) through (f)]. Although Sec. 210-52(c) requires the installation of "countertop" receptacles on closer spacings than the other receptacles required by Sec. 210-52(a), it *does not* in any way modify the requirement given by the last paragraph of Sec. 210-52(a). Therefore, a receptacle that is part of a lighting fixture may *not* serve to satisfy the requirement for countertop receptacles as given in Sec. 210-52(c).

The answer to the second question regarding circuiting and GFCI protection (where the receptacle is within 6 ft. of a kitchen sink) is yes.

Sec. 210-52(b) clearly requires *all* receptacle outlets in the "kitchen, pantry, breakfast room, dining room, or similar area" to be installed on the two or more small appliance branch circuits required by Sec. 220-4(b). In addition, Sec. 210-52(b) states that the two or more small appliance branch circuits must supply receptacles, outlets, and *only* receptacle outlets. Therefore, to comply with the literal wording of Sec. 210-52(b), the receptacle within the lighting fixture in question must be supplied from one of the two or more small appliance branch circuits. And the lighting fixture itself must be supplied from a circuit *other than* one of the two or more small appliance branch circuits.

Sec. 210-8(a) covers "Ground-Fault Circuit-Interrupter Protection for Personnel" in dwelling units. Part (5) of Sec. 210-8(a) calls for *all* 15 and 20A, 125VAC receptacles that are installed within 6 ft. of *any* kitchen sink to be provided with GFCI protection. No exception is made for receptacles that are part of a lighting fixture or appliance. Therefore, compliance with the literal wording of Sec. 210-8(a)(5) can only be achieved by providing GFCI protection for the lighting fixture receptacle where installed within 6 ft. of a kitchen sink.

It is worth noting that in today's modern kitchens, there may be more than one sink and GFCI protection must be provided for every receptacle that is within 6 ft. of *any* sink in the kitchen of a dwelling unit. Although not required by the literal wording of Sec. 210-8(a)(5), voluntarily providing GFCI protection for *any* receptacle located within 6 ft. of *any* sink in *any* room of a dwelling unit will assure the same level of safety for people in those areas and rooms that is required for receptacles at kitchen (and bathroom) sinks. If the benefits of such practice is explained to the home owner, I believe that most home owners will welcome the suggestion *and* be willing to pay for the additional measure of safety provided.

Permission to Reidentify Conductors Is Limited

Secs. 210-5(b), 250-57(b), 310-12(b), and 517-13(a), Ex. No. 1. Use of non-green-colored insulated conductors as an equipment grounding conductor is only permitted where "qualified persons" will maintain the installation.

As is the case with any new edition of the Code, it takes from a year to a year and a half for the majority of jurisdictions to adopt the new Code. The 1990 National Electrical Code is no different. One of the more interesting aspects of this proliferation of acceptance is the many and varied ways in which electrical professionals involved in the design and installation of electrical systems and equipment interpret the rules of the "new" Code and the variety of actual methods used to achieve compliance with the literal wording.

In some cases, a change in the wording of a rule will cause engineers and contractors to completely rethink their approach to the application. Many times the results are a "new" method that will satisfy the "new" rule and at the same time offer some benefit—typically an economic benefit—to the designer, contractor, and/or customer.

One such "new" method for complying with a change in the 1990 NEC is related to the change in Sec. 517-13(a), Exception No. 1.

In the never-ending pursuit to remain competitive during the bid process and at the same time comply with this "new" Code rule, a number of contractors have again shown that the ingenuity of the American electrician is never to be underestimated. Instead of using a

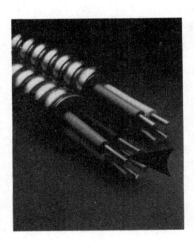

"premium priced" Type AC cable that is specifically manufactured with a green insulated equipment grounding conductor for wiring fixed equipment and receptacles in patient-care areas of health-care facilities, these individuals are using a "standard" Type AC cable—with 3 or 4 insulated conductors—and are reidentifying one of the insulated copper conductors as the equipment grounding conductor at the time of installation with green tape or by coloring the conductor insulation green.

The question is: Does such an application satisfy all applicable Code rules or does it expose the installer (or designer, if reidentification is specified) to possible legal action in the event of an accident?

Analysis

The first consideration relates to whether or not such an application satisfies the rules of Article 517. Sec. 517-12 states that unless a specific rule of Article 517 modifies the general requirements given in Chapters 1 through 4 of the NEC, then all wiring within health-care facilities must satisfy the rules given in Chapters 1 through 4.

Before one can determine whether or not the rules of Article 517 modify or otherwise amend the requirements given in Chapters 1 through 4 for identification of the equipment grounding conductor, it is first necessary to look at the general requirements given in Chapters 1 through 4.

The NEC provides guidance on identification of equipment grounding conductors in three different sections: Secs. 210-5(b), 250-57(b), and 310-12(b). (The wording of the applicable rules of Sec. 250-57(b) is shown in Figure 1.)

The basic rule of each of these sections states that an equipment grounding conductor may be bare, covered, or insulated. And an individually covered or insulated equipment grounding conductor must be colored "green, or green with one or more yellow stripes" for its entire length (Figure 2).

Exception No. 1 to Sec. 210-5(b), which covers identification of equipment grounding conductors in branch circuits, recognizes the identification methods described in Exception Nos. 1 and 3 to Sec. 250-57(b) and Exception Nos. 1 and 2 to Sec. 310-12(b). The rules that cover reidentification of an insulated conductor in a multiconductor cable are Sec. 250-57(b), Exception No. 3 and 310-12(b), Exception No. 2. These two exceptions are nearly identical and an examination of Exception No. 3 to Sec. 250-57(b) will serve to clarify what is required by both rules.

Exception No. 3 to Sec. 250-57(b) [and Exception No. 2 to Sec. 310-12(b)] recognizes reidentification of an insulated conductor as the

250-57. Equipment Fastened in Place or Connected by Permanent Wiring Methods (Fixed) — Grounding. Noncurrent-carrying metal parts of equipment, raceways, and other enclosures, where required to be grounded, shall be grounded by one of the methods indicated in (a) or (b) below.

(b) With Circuit Conductors. By an equipment grounding conductor contained within the same raceway, cable, or cord or otherwise run with the circuit conductors. Bare, covered or insulated equipment grounding conductors shall be permitted. Individually covered or insulated equipment grounding conductors shall have a continuous outer finish that is either green, or green with one or more yellow stripes.

Exception No. 3: Where the conditions of maintenance and supervision assure that only qualified persons will service the installation, an insulated conductor in a multiconductor cable shall, at the time of installation, be permitted to be permanently identified as an equipment grounding conductor at each end and at every point where the conductor is accessible by one of the following means:

a. Stripping the insulation from the entire exposed length,

b. Coloring the exposed insulation green, or

c. Marking the exposed insulation with green tape or green colored adhesive labels.

Figure 1   Portions of Sec. 250-57 that give the Code requirements for identification of equipment grounding conductors in multiconductor cables. These rules are nearly identical to those given in Secs. 210-5(b), 310-12(b), and their applicable Exceptions. Although the NEC *does* recognize reidentifying an insulated conductor of *any* size in a multiconductor cable as the equipment grounding conductor at the time of installation, such permission may *only* be exercised under the conditions put forth in the first sentence of Exception No. 3.

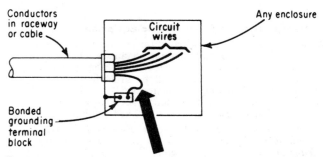

Equipment grounding conductor may be bare, or covered to show a green color or green with one or more yellow stripes.

Figure 2   The basic Code rule for identification of an equipment grounding conductor. Whenever the grounding conductor is not bare, or an insulated or covered conductor is not green or green with one or more yellow stripes for its entire length, then one of the methods described in the exceptions to the basic rule must be used. Although these requirements are covered in a number of sections, this matter is covered most comprehensively in Sec. 250-57(b).

equipment grounding conductor at the time of installation where none of the insulated conductors in a multiconductor cable is colored green or green with one or more yellow stripes as required by the basic rule. This reidentification may be accomplished by:

1. Stripping the insulation or covering back to the point where the conductor leaves the raceway and enters the enclosure (switchboard, panelboard, junction box, pull box, device box, etc.), or

2. Coloring the end of the conductor insulation or covering "green" at the time of installation, or

3. Marking the end of the conductor insulation or covering with green-colored tape or labels at the time of installation.

And the equipment grounding conductor must be reidentified at every point in the system where the grounding conductor is accessible.

With the preceding information in mind, a comparison between the requirements of Sec. 250-57(b) and its Exception No. 3 (Figure 1) and the rules of 517-13(a) and its Exception (Figure 3) reveals that:

1. According to the rule of Sec. 517-13(a), the equipment grounding conductor does *not* have to be "green or green with one or more yellow stripes" for its entire length. This rule simply states that the required equipment grounding conductor be an *insulated* conductor.

2. Although recognized by Exception No. 3 to Sec. 250-57(b) and Exception No. 2 to 310-12(b), stripping of the insulation would not be

---

517-13. Grounding of Receptacles and Fixed Electric Equipment.

(a) Patient Care Areas.   In areas used for patient care, the grounding terminals of all receptacles and all noncurrent-carrying conductive surfaces of fixed electric equipment likely to become energized that are subject to personal contact, operating at over 100 volts, shall be grounded by an insulated copper conductor. The grounding conductor shall be sized in accordance with Table 250-95 and installed in metal raceways with the branch-circuit conductors supplying these receptacles or fixed equipment.

Exception No. 1: Metal raceways shall not be required where Type MI cable, and Types MC and AC cables where the outer metal jacket is an approved grounding means of a listed cable assembly.

Figure 3 Sec. 517-13(a) and its Exception. Although the requirement for an insulated copper equipment grounding conductor was dropped from the revised wording of the Exception to this rule, where Type MI, Type MC, or Type AC cable is used in accordance with this Exception, the cable is still required to contain an *insulated copper* equipment grounding conductor. The Exception is only intended to eliminate the need for a metallic raceway, as called for in the basic rule, but the remainder of the basic rule (including providing an insulated copper grounding conductor) must be satisfied.

permitted for wiring in a patient-care area of a health-care facility because the rule of Sec. 517-13(a) specifically requires the equipment grounding conductor to be *insulated*.

3. Sec. 517-13(a) also requires the equipment grounding conductor to be *copper*. Although aluminum or copper-clad aluminum conductors are generally recognized for use as an equipment grounding conductor, use of an aluminum or copper-clad aluminum conductor as the grounding conductor in patient-care areas of a health-care facility would be in direct violation of the wording given in the basic rule of Sec. 517-13(a).

Put simply, reidentification of an insulated conductor as the equipment grounding conductor in a multiconductor cable would seem to comply with both Exception No. 3 to Sec. 250-57(b) and the Exception to Sec. 517-13(a) and should be acceptable for supplying fixed equipment or receptacles in patient-care areas of a health-care facility if an *insulated copper* conductor is reidentified at the time of installation by green coloring or taping/labeling on the insulation.

But have we fully satisfied the requirements of Sec. 250-57(b), Exception No. 3? The answer is no.

There is one more important (and often overlooked) qualification that must be satisfied before exercising the permission given for reidentification of an insulated conductor as an equipment grounding conductor in a multiconductor cable. Such permission may be used only under the conditions given in the first two lines of Sec. 250-57(b), Exception No. 3. That is, a conductor in a multiconductor cable may be identified as the equipment grounding conductor by green-coloring or -taping when wiring in a patient-care area of a health-care facility, *but only* where "conditions of maintenance and supervision assure that only qualified persons will service the installation." What does this mean and how does the designer or installer determine if a particular installation meets this requirement?

Some guidance is provided by the original Code-change proposals and Panel Action that appeared in the NFPA's "Preprint of the Proposed Amendments for the 1978 National Electrical Code," which was the forerunner of today's "Technical Committee Report" (TCR). One of the two nearly identical proposals to add an Exception No. 3 to Sec. 250-57(b) and the Panel's proposed wording are reproduced in Figure 4.

As can be seen, the wording of the submitted proposal is not at all concerned with who maintains or services the installation after the fact. Both proposals simply sought relief from the basic color-coding requirement for equipment grounding conductors where multiconductor cables were used.

We can also see the original wording proposed by the Code Making Panel (CMP) limited such permission to "industrial and mining appli-

250-57(b) Exception No. 3—(New): Reject

PROPOSAL NO. 71: Add Exception No. 3 as follows:

Exception No. 3: Where wiring employs multiconductor cables, identification of the grounding conductor by numbering or lettering shall be permitted. See Section 310-10(a) Exception No. 5 and Section 310-10(b) Exception No. 2.

SUBMITTER: IEEE

SUPPORTING COMMENT: The rationale for requiring coding of grounded conductors by colors is excellent for single conductors in the raceway types for which they are suited. In multiconductor cable, however, the several other types of identification, particularly numbering and lettering have proven amply satisfactory and safe. They have an additional advantage in that the human handicap of color-blindness does not cause errors and associated hazards. Too, the necessity of applying standard multiconductor cables to a multiplicity of circuit applications, only a few of which involve grounded conductors, would impose severe burdens which are not warranted in view of the freedom from problems using the proposed systems.

PANEL RECOMMENDATION: Reject

PANEL COMMENT: See Proposal 71-1.

VOTE ON PANEL RECOMMENDATION: Unanimously affirmative

Figure 4   Reproduction from the 1978 "Preprint" of one of two nearly identical proposals (Proposals No. 71 and 72) submitted for the 1978 NEC that would recognize methods other than the color-coding requirement given in the basic rule for identification of the equipment grounding conductor in multiconductor cables of any size (above). Although the CMP did accept the concept, they changed the wording of the proposal (as can be seen opposite). The wording that now appears in the 1990 NEC was what finally appeared in the 1978 NEC in Sec. 250-57(b), Exception No. 3, although there is no definite indication in available documentation as to why the wording was amended from the time the "Preprint" was published and the accepted final version of the 1978 NEC was published.

cations." The final wording of the rule as it appeared in the 1978 NEC, however, did not limit application to "industrial and mining" facilities, but rather stated that such permission may only be exercised where "conditions of maintenance and supervision assure that only qualified persons will service the installation."

Because there is no additional documentation available that might further clarify what happened between the time the CMP accepted the original wording and the time that the 1978 Code was published using the same wording that still appears in the 1990 edition, it seems reasonable to conclude that the CMP did not want to limit this permission to only "industrial and mining applications," and they also did not want it used on a wholesale basis. From the initial and final wording used in this rule, it would seem that such permission may be used in *any* building, structure, facility, institution, etc., provided the building, structure, etc., has a competent in-house or outside electrical work force to maintain and service the installation.

250-57(b) Exception No. 3—(New): Accept

PROPOSAL NO. 71-1: Add Exception No. 3 to Section 250-57(b) as follows:

Exception No. 3: In industrial and mining applications an insulated grounding conductor in a multiconductor cable shall, at the time of installation, be permitted to be suitably identified as a grounding conductor at each end and at every point where the conductor is accessible. Identification shall be accomplished by one of the following:

a. Stripping the insulation from the entire exposed length,

b. Coloring the exposed insulation green, or

c. Marking the exposed insulation with green tape or green colored adhesive labels.

SUBMITTER: CMP5

SUPPORTING COMMENT: To accomplish intent of proposals 71 and 72 but to require that the reidentification means is basically bare or green.

PANEL RECOMMENDATION: Accept

VOTE ON PANEL RECOMMENDATION: Unanimously affirmative

COMMENT ON VOTE:

FARQUHAR: There are two aspects of this proposal which concern us. We believe that a conductor which is insulated or covered with a continuous outer finish which is either green or green with one or more yellow stripes should not be permitted to be identified and used for any other purpose. The color, green, should be reserved for use only on grounding conductors.

We further believe that any conductor which is not provided with a green covering and which is identified at the time of installation as a grounding conductor should be required to be so identified by one of the means called for in Section 310-10(b) Exception and not by the simple use of numbers, letters, or other similar marking.

Figure 4   (Continued).

Conclusion

Although most major health-care facilities do have a competent electrical work force (either in-house or contracted from outside), they're not the ones generally employing this technique. It is in the smaller-type health-care facilities—such as dental and medical offices located in the doctor's residence, small clinics and outpatient treatment centers, and other smallish "health-care facilities" (as defined in Article 517)—that this approach has gained popularity. And, in many cases, these facilities have no qualified electrical maintenance personnel, either on staff or contracted from outside. In these health-care facilities, reidentification of an insulated conductor as the equipment grounding conductor in a multiconductor cable actually *violates* the permission given in Sec. 250-57(b), Exception No. 3 and should *not* be used.

If you are not absolutely *positive* that *only* qualified personnel will service the installation—presumably for the entire life of the installation—at a health-care facility (or *any* other building, structure, or facility for that matter), then *do not* exercise the permission given by the rule of Sec. 250-57(b), Exception No. 3. Or, better still, always comply with the basic rule and completely avoid leaving yourself open to any question at a future date regarding how you determined that only qualified electrical maintenance personnel would service and maintain the installation for the life of the installation. The same approach should be applied to any other Code rule that presents the same or a similar qualifying condition, such as Sec. 200-6(a) Exception No. 3, 200-6(b) Exception, 200-9 Exception, etc.

Whenever an exception presents a qualifying condition that is not clear or well defined, as a general rule try to follow the basic requirement and ignore the exception. In this case, use a Type AC cable that contains a conductor with an overall green coloring or an overall green coloring with one or more yellow stripes and forget about whether or not the installation will be serviced and maintained by competent electrical personnel. The additional cost should be viewed as "cheap" insurance that will definitely limit legal exposure for the designer and/or installer should an accident occur. But from a "real life" standpoint, the far more important argument for using a listed, manufactured cable with a fully green-colored equipment grounding conductor is the elimination of the very expensive labor costs that are involved in doing all the reidentification. That labor cost should more than offset any differential in material costs. And you're absolutely positive that the grounding conductor is properly identified.

Circuits for Receptacles

Secs. 210-50, 210-52, 210-63, 220-2, 220-3, and 220-4.   Circuits required for the supply of convenience receptacle outlets varies according to the type of installation.

Design of branch circuits for plug receptacles requires careful determination of particular requirements. The type and size of occupancy and nature of the work performed there will indicate the best manner of handling plug-connected loads. Of particularly important concern today is the circuiting of receptacle circuits that are used in office areas to supply heavy concentrations of personal computers, printers, and other business equipment that is powered by switching-mode power supplies. The concern in all such cases is to properly and effectively account for high levels of harmonic currents on the circuit neutrals.

For the maximum convenience of a building's tenants or owners, modern design provides for plug-in devices to be supplied from receptacles on circuits other than general lighting circuits. In this way, the loading of circuits can be kept under control, and spare capacity can be more realistic.

Code limitations on the number of plug outlets permitted on any one branch circuit should be carefully observed in selecting the number of receptacle circuits.

Nonresidential occupancies

In Sec. 210-50, the Code specifies where and when receptacle outlets are required on branch circuits. Note that there are no general or specific requirements for receptacle outlets in commercial, industrial, and institutional installations other than for store show windows. In Sec. 210-50(b), there is the general rule that receptacles do have to be installed where flexible cords with plug-caps are used. In nonresidential buildings, if flexible cords with plug-caps are not used, there is no requirement for receptacle outlets. They have to be installed only where they are needed, and the number and spacing of receptacles (in other than dwelling units) are completely up to the designer. But because the Code takes the position that receptacles in nonresidential buildings only have to be installed where needed for connection of specific flexible cords and caps, it demands that where such receptacles are installed, each must be taken as a load of 180VA.

Sec. 220-3(c)(5) requires that every general-purpose, single, duplex, or triplex convenience receptacle outlet be taken as a load of 180VA, and that amount of circuit capacity must be provided for each such outlet. This rule calls for "each single or each multiple receptacle *on one strap*" to be taken as a load of "not less than 180 voltamperes"—in commercial, institutional, and industrial occupancies. Each individual device strap—whether it holds one, two, or three receptacles—is a load of at least 180VA (Figure 1).

If a 15A, 115V circuit is used to supply *only* receptacle outlets, then the maximum number of general-purpose receptacle outlets that may be fed by that circuit is 15A times 115V divided by 180VA or *9 receptacle outlets.*

For a 20A, 115V circuit, the maximum number of general-purpose receptacle outlets is 20A times 115V divided by 180VA or *12 receptacle outlets.*

Note: *In these calculations, the actual results work out to be 9.58 receptacles on a 15A circuit and 12.77 on 20A circuits.*

The value of 115V is used in these calculations to enhance adequacy, although in a reference to "voltage" covering the sample calculations in Chapter 9 in the back of the book, the NEC does specify:

For uniform applications of Articles 210, 215, and 220, a nominal voltage of 120, 120/240, 240 and 208Y/120V *shall be* used in computing the ampere load on conductor.

In the above calculations of number of receptacles permitted on a circuit, if 120V is used instead of 115V, then a 15A circuit has a voltampere capacity of 1800VA and dividing that by 180VA permits 10 receptacles on a 15A circuit. A 20A circuit could have 13 receptacles. The NEC permits that number of receptacles on each circuit.

Sec. 220-2 also *requires* that the 120 and 240 values of voltage be used in making electrical calculations to determine conductor ampacity for branch circuits and feeders. Although both of those Code rules use the word "shall" and seem to make it mandatory that the higher values of voltage be used in making calculations, use of the lower voltage values (115 and 230V instead of 120 and 240) would result in higher current values when voltage is divided into load voltamperes and would, therefore, dictate use of larger conductor sizes—which could not be a Code violation. Using 115 and 230V instead of 120 and 240 does produce higher current values when dividing volts into watts or voltamperes (e.g., 6000W divided by 120V equals 50A; 6000W divided by 115V equals 52A). Using higher cur-

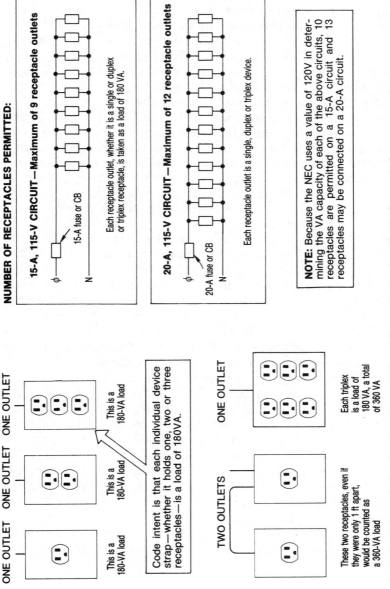

NUMBER OF RECEPTACLES PERMITTED:

15-A, 115-V CIRCUIT—Maximum of 9 receptacle outlets

15-A fuse or CB

Each receptacle outlet, whether it is a single or duplex or triplex receptacle, is taken as a load of 180 VA.

20-A, 115-V CIRCUIT—Maximum of 12 receptacle outlets

20-A fuse or CB

Each receptacle outlet is a single, duplex or triplex device.

NOTE: Because the NEC uses a value of 120V in determining the VA capacity of each of the above circuits, 10 receptacles are permitted on a 15-A circuit and 13 receptacles may be connected on a 20-A circuit.

ONE OUTLET   ONE OUTLET   ONE OUTLET

This is a 180-VA load

This is a 180-VA load

This is a 180-VA load

Code intent is that each individual device strap—whether it holds one, two or three receptacles—is a load of 180VA.

ONE OUTLET

Each triplex is a load of 180 VA, a total of 360 VA

TWO OUTLETS

These two receptacles, even if they were only 1 ft apart, would be counted as a 360-VA load

Figure 1 The number of general-purpose receptacles permitted to be connected on a single circuit is limited for applications in nonresidential (industrial, commercial, institutional) facilities.

47

rent values for sizing conductors and other equipment gives greater assurance of adequate capacity and satisfies all NEC objectives. In calculations for number of receptacles, use of the lower voltages enhances adequacy.

Based on the very clear experience and evidence that electrical systems quickly tend to be overloaded and because load growth is a certainty in this all-electric age, many designers resolve all questions about circuit loading and permitted capacity in favor of greater capacity. Use of 9 instead of 10 as the maximum number of receptacles on a circuit calls for more circuits and gives higher capacity to each receptacle outlet. "When in doubt, design up, not down."

For offices and buildings and banks, the NEC permits either of two methods of determining required minimum branch-circuit capacity for receptacle outlets. The first is the method just described in which a load of 180VA is taken for each single or duplex receptacle outlet and sufficient branch-circuit capacity is provided for the total load (180VA × the number of receptacle outlets). The second—or alternative approach—is described in the last footnote to NEC Table 220-3(b) (the double-asterisk note).

Based on extensive analysis of load densities for general lighting in office buildings, NEC Table 220-3(b) requires a minimum load unit of 3-1/2VA/sq. ft. for general illumination in "office buildings" and for "banks." The minimum voltamperes of branch-circuit capacity for such occupancies must not be less than 3-1/2 × the sq. ft. floor area of the space being supplied with the general lighting. But the double asterisk note at the bottom of the table requires the addition of another 1VA/sq. ft. to the 3-1/2VA/sq. ft. value to cover the loading added by general-purpose receptacles *in those cases* where the actual number of receptacles is not known at the time feeder branch-circuit capacities are being calculated. In such cases, a unit load of 4-1/2VA/sq. ft. must be used; and the calculation based on that figure will yield an NEC acceptable value of minimum branch-circuit capacity for both general lighting and for all general-purpose receptacles that may later be installed. Whatever the number of receptacles that may later be installed, they may be divided among lighting circuits or separate receptacle circuits, with the loads evenly proportioned among the circuits. In such cases, the required allowance of 180VA for each receptacle outlet is simply disregarded.

But the rule does require that, where the actual number of general-purpose receptacles is known, the general lighting load is taken at

3-1/2VA/sq. ft. for branch-circuit and feeder capacity, and each receptacle is taken as a load of at least 180VA to get the total required branch-circuit capacity. The demand factors of Table 220-13 may be applied to get the minimum required feeder capacity for receptacle loads.

Exception No. 1 to Sec. 220-3(c) requires branch-circuit capacity to be calculated at 180VA for each 5-ft. length of multioutlet assemblies (pre-wired surface metal raceways with plug outlets spaced along their length). But, each 1-ft. length of such strip must be taken as a load of 180VA when the strip is used where a number of plug-in appliances are likely to be used simultaneously. For instance, in the case of industrial applications on assembly lines involving frequent, simultaneous use of plugged-in tools, the loading of 180VA/ft. must be used. (A loading of 180VA for each 5-ft. section may be used in commercial, industrial, or institutional applications of multioutlet assemblies when the use of plug-in tools or appliances is not heavy.)

Residential occupancies

Although the above-described Code rules cover the maximum permitted number of receptacle outlets in commercial, industrial, institutional, and other nonresidential installations, there are no such limitations on the number of receptacle outlets on residential branch circuits.

A different approach is used for receptacles in dwelling-type occupancies. The Code simply assumes that cord-connected appliances will always be used in all residential buildings and requires general-purpose receptacle outlets of the number and spacing required by Secs. 210-52 and 210-60. These rules cover one-family houses, apartments in multi-family houses, guest rooms in hotels and motels, living quarters in dormitories, etc. But because so many receptacle outlets are required in such occupancies and because use of plug-connected loads is intermittent and has great diversity of load values and operating cycles, the Code notes at the bottom of Table 220-3(b) that the loads connected to such receptacles are adequately served by the branch-circuit capacity that is provided for general lighting, as required by Sec. 220-4, and no additional load calculations are required for such outlets.

In dwelling occupancies, it is first necessary to calculate the total "general lighting load" from Sec. 220-3(b) and Table 220-3(b) (at 3W/sq. ft.

for dwellings or 2W/sq. ft. for hotels and motels, including apartment houses without provisions for cooking by tenants) and then provide the minimum required number and rating of 15A and/or 20A general-purpose branch circuits to handle that load as covered in Sec. 220-4(a). As long as the basic circuit capacity is provided, any number of lighting outlets may be connected to any general-purpose branch circuit, up to the rating of the branch circuit if loads are known. The lighting outlets should be evenly distributed among all of the circuits. Although residential lamp wattages cannot be anticipated, the Code method covers fairly heavy loading.

When the above Code rules on circuits and outlets for general lighting are satisfied, general-purpose convenience receptacle outlets may be connected on one or more of the required branch circuits; or additional circuits (over and above those required by the code) may be used to supply the receptacles. But no matter how general-purpose receptacle outlets are circuited, *any number of general-purpose receptacle outlets may be connected on a residential branch circuit—with or without lighting outlets on the same circuit.*

In addition, when small-appliance branch circuits (in kitchen, dining room, pantry, etc.) are provided in accordance with the requirements of Sec. 220-4(b), *any number* of small-appliance receptacle outlets may be connected on the 20A small-appliance circuits—but only receptacle outlets may be connected to these circuits and only in the specified rooms.

Sec. 210-52(a) applies to spacing of receptacles connected on the 20A small-appliance circuits, as well as spacing of general-purpose receptacle outlets. That section, in effect, establishes the *minimum* number of receptacles that must be installed. Of course, a greater number of receptacles may be installed for greater convenience of use.

Receptacle circuits/harmonics

Over recent years, electrical circuits supplying receptacles for data processing equipment have been plagued with severe overheating of branch-circuit neutral conductors due to harmonic currents that are *additive* in the neutral of 3-phase, 4-wire circuits supplying such loads. This problem has been widely experienced, often with dangerous and destructive effects. Experience has shown that many 3-phase, 4-wire circuits supplying balanced phase loading of computer equipment have neutral harmonic currents with RMS values considerably greater than the phase currents—up to 2 or more times the phase current. The actual level of harmonic neutral current will depend on the load devices themselves. Severe overheating of neutral conductors results from the conductors being subjected to load currents in excess of their ampacities.

The overheating of such neutral conductors arises, of course, from very high I^2R heat losses in the conductor when the conductor is carrying current in excess of its NEC ampacity, producing temperatures that exceed the insulation temperature rating (e.g., 75°C of THW, etc.). As a general attempt to limit and control heating of harmonic current loading on neutrals, a larger than usual neutral conductor can be selected to reduce the resistance of the conductor as a means of reducing heating. Some designers specify use of a neutral conductor at least one size larger than required phase conductors for a 3-phase, 4-wire branch circuit supplying computer/data processing equipment. Others may go up two sizes. This is a difficult problem because circuit design is generally done well in advance of any firm knowledge of the relative offending nature (harmonic generation) of equipment, which will be selected and applied at a later date.

Another heat-control technique that has been widely utilized is providing a separate neutral conductor for each phase leg of the circuit. So, instead of being a 3-phase, 4-wire circuit (three phase legs and one neutral), some circuits are designed as 3-phase, 6-wire branch circuits—based on the concept that in a 3-phase, 6-wire circuit, each neutral will carry only 1/3 the amount of harmonic current that would flow on a single common neutral of a 3-phase, 4-wire circuit. If a single, common neutral conductor of 0.1 ohm resistance for its entire length carried 15A of harmonic current (5A per phase leg), the heating effect (I^2R) would be $15 \times 15 \times 0.1 = 0.1 = 22.5$ watts. But if the 15A of harmonic current are divided among three neutral conductors of 0.1 ohm resistance each, then each conductor would have a heat loss (I_2R) of $5 \times 5 \times 0.1 = 2.5$ watts. And the three neutrals together would have a heat loss of 3×2.5 watts, or 7.5 watts instead of 22.5.

But, when a separate neutral is used for each phase leg of the branch circuit, there is an entirely different condition with respect to the 60Hz fundamental current—because the 60Hz neutral current is not cancelled, as in the 3-phase, 4-wire circuit. Then a question arises as to the relative heating effects of the 60Hz current *plus* the harmonic currents as they both flow in the separate neutrals—compared to the heating effect in a common neutral, which carries *only* the harmonic current (because the 60Hz currents cancel to zero in a common neutral).

Figure 2 shows the comparison of heating effects for the two different circuit makeups.

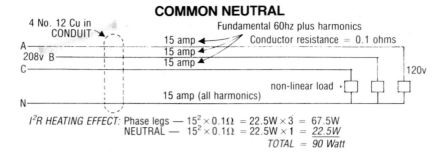

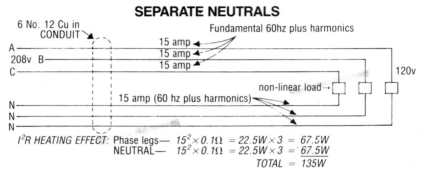

Figure 2 Comparing a "common neutral," 3-phase, 4-wire receptacle circuit (at top) with the same circuit using a separate neutral for each hot leg shows that the common-neutral circuit has lower losses for low levels of harmonic loading—such as here where each phase leg has, say, 10 amps of fundamental current and 5 amps of harmonics. But if the phase loading was 10 amps harmonics and only 5 amps fundamental, the "common-neutral" circuit at top would have an additive 30 amps on the neutral—for a NEUTRAL loss of $30^2 \times 0.1 = 900 \times 0.1 = 90$ watts. In such a case, the NEUTRAL loss of 67.5 watts for the separate-neutral circuit at bottom would dictate use of separate neutrals for the hot legs. But a single larger neutral, with an "R" lower than 0.1 ohms, would have to be considered.

The "Common Neutral" circuit with a larger neutral (say, No. 10 or No. 8) provides the most effective and economical reduction of heating in circuits with small to moderate amounts of harmonic content. As harmonic current loading becomes heavy, very large increases in resistance due to "skin effect" and "proximity effect" would dictate use of "Separate Neutrals"—especially where the ratio of harmonic to fundamental is very high.

USE A SINGLE, COMMON NEUTRAL ALL THE TIME?

A single, larger common neutral—such as one No. 8 neutral with three No. 12 phase conductors—can offer more advantage than separate neutrals, even with circuits where the ratio of harmonic to fundamental is very high. As discussed in the caption for Fig. 40, with circuit loading of 10A harmonic and 5A fundamental, the common neutral loss would be 90 watts vs 67.5 watts loss in three neutrals. But that is for No. 12 neutral conductors in both cases. Using a single No. 8 common neutral, for a circuit length that gives 0.1 ohms for No. 12, the No. 8 would have a resistance of only 0.039 ohms. The I^2R neutral loss for the No. 8 would be 30^2 x 0.039 = 900 x 0.039 = 35 watts. That is considerably less than the 67.5 watts for the three separate neutrals. If the single, larger neutral was No. 10 instead of No. 8, neutral loss would be about the same for either one No. 10 common neutral or three No. 12 separate neutrals.

CONCLUSION: For the 3-phase part (the "home-run," at least) of any receptacle circuit to nonlinear load devices, a single neutral of increased size is more efficient—as well as more economical in labor and material—than separate neutrals.

Required Outlets at Heating, Air-Conditioning, and Refrigeration Equipment

Secs. 210-52, 210-63, and 210-70(a). Use of fixtures with integral receptacle may *not* satisfy NEC requirements for maintenance receptacle at heating, air-conditioning, and refrigeration equipment.

Question: An inspector recently turned down a job where I had installed a pull-chain porcelain lighting outlet with an integral receptacle outlet as the required lighting and receptacle outlet for an air-conditioning/heat/air-handler unit that was installed in the attic of a home I was wiring.

A wall-mounted switch that controls the pull-chain porcelain lampholder was already installed at the point of entrance to the attic, complying with the rule of Sec. 210-70(a).

The inspector stated that a separate box and individual circuit was required for the receptacle. And that the lighting outlet was to be "keyless" and not of the pull-chain type.

As can be seen in this photo, the receptacle within the pull-chain porcelain lampholder is rated for 15A at 125V. The wording of Sec. 210-63 does not specifically require that the receptacle be on a separate circuit nor does it prohibit switching of the receptacle.

My question is: Would the above pictured combination pull-chain lampholder and receptacle satisfy the requirement of Sec. 210-63 or must the receptacle be installed on a separate circuit and in a separate box as indicated by the inspector?

Answer: This question has come up at many Code-change meetings in various areas of the country. It is a good question and one that is not easily answered because there are several aspects that are not directly addressed by the Code rules involved.

As you stated, there is no specific requirement in Sec. 210-63 for the required "receptacle outlet" to be installed on a separate or dedicated circuit. And there is no prohibition against considering a receptacle within a lighting fixture as the "required receptacle." Even though a receptacle within a lighting fixture is generally not permitted to be considered a "required receptacle" by the rule of Sec. 210-52, this rule appears to apply only to the receptacles required by that Code section, and *not* the receptacle required by Sec. 210-63.

With this in mind, it would seem that such a device does satisfy the requirements of Sec. 210-63. However, there is another sticky point in the wording of Sec. 210-63 that presents a problem. Does the receptacle within the lampholder comply with the requirement for a "receptacle outlet"?

As given in Article 100 of the 1990 NEC, a "receptacle outlet" is: *An outlet where one or more receptacles are installed.* This definition can be interpreted to mean that a receptacle outlet is an outlet where *only* receptacles—one or more—are installed. If the local electrical inspector interprets the term "receptacle outlet" to mean *only* receptacles may be installed, then use of the device in question would *not* satisfy the wording in Sec. 210-63. In this case, this is the way the rule should be interpreted. I, personally, would not accept the use of a receptacle in a lighting fixture as satisfying the rule of Sec. 210-63.

The only rule that addresses "switching" of the required receptacle is the prohibition against supplying the receptacle from the load-side of the equipment disconnecting means. The idea behind this rule is that when somebody opens the equipment disconnect for the air-conditioning, heating, or refrigeration equipment prior to servicing, the receptacle, which is intended to provide local power for electric tools and/or test equipment and thereby eliminate the need for long extension cords, would be "dead" and not capable of performing its desired function.

In this case, because the receptacle is switched with the lighting outlet by the wall switch at the point of entrance, it would seem that there would be no compromise to the intent of this rule. Therefore, connection of the required receptacle to the same circuit as the lighting fixture and switching of the receptacle along with the lighting fixture should not be objectionable, whether the receptacle is in an individual outlet box or if it is part of the lighting fixture.

As can now be seen, this is not an application that can be simply answered yes or no, based on the wording of the Code rules involved. And, as always, the local electrical inspector has responsibility for "approving" equipment and materials, as well as interpreting Code rules (see Sec. 90-4). Therefore, it is the local electrical inspector who has the final say.

Service Load: Continuous or Noncontinuous?

Secs. 220-10(b) and 310-15. Rating of service overcurrent device must be equal to the noncontinuous load *plus* 125 percent of the continuous load.

Question: After determining service load in accordance with Art. 220, is the calculated load considered to be a "continuous" load or is that value a "noncontinuous" load? I mean, is it necessary to multiply the calculated current value by 1.25 to determine the rating of the service overcurrent protective device or is it acceptable to use a non-100 percent-rated CB or fuses whose ampere rating is equal to the calculated load?

Answer: This is a good question. The service (or feeder) load calculated in accordance with Art. 220 is the minimum amount of current carrying capacity (ampacity) required for the *conductors* supplying the load. That is, the ampere value determined in accordance with the requirements of Art. 220 indicates how much current the service (or feeder) conductors will see at any given time. As recognized by the various demand factors given in Art. 220, the total connected load will virtually never be carried by the service conductors and, therefore, the service conductors may have an ampacity less than the total connected load.

As far as the *service conductors* are concerned, Sec. 220-10 states that the service or feeder conductors must have "sufficient ampacity to supply the load being served." From the definition in Art. 100, we know that ampacity *is* a *continuous* current rating. That is, ampacity as determined in accordance with Sec. 310-15, will be the amount of current, in amperes, that the conductor is capable of carrying *continuously*.

However, to determine if the service load, as calculated in accordance with Art. 220, is continuous or noncontinuous, the individual loads must be evaluated with respect to the definition for "continuous load," as given in Art. 100. If the load is energized for 3 hours or more, then it is considered to be "continuous." If not, then the load is "noncontinuous." For dwellings there are very few loads that would be considered "continuous": outdoor lighting, certain types of storage heating systems, pool pumps, permanently installed fans, and other equipment that would be operated for a period of 3 hours or more. (Notice that ovens are not included. Although an oven may be on for, say, 6 hours while you are cooking a Thanksgiving turkey, the thermostat will switch the heating elements on and off based on the internal oven temperature. Although the oven switch may be in the ON position, the oven will not actually draw current on a continuous basis.)

Although "continuous" versus "noncontinuous" load is not a concern for the conductors, the overcurrent protective device must be rated for the combined value of "noncontinuous" current *plus* 1.25 times the "continuous" current, unless the CB or fused switch is marked as suitable for loading to 100 percent of its rating. [See Secs. 210-22(c), 220-3(a), 220-10(b), and 384-16(c)].

One method to assure adequacy of a selected protective device is to take all continuous load at 125 percent of its ampere rating when determining conductor size. Then when an overcurrent protective device is selected to protect that conductor, the rating of the protective device will be equal to, or greater than, the noncontinuous plus 1.25 times the continuous load. For example:

$$\text{Service load (per Art. 220)} = 130A$$

$$\text{Noncontinuous load} = 84A$$

$$\text{Continuous load} = 46A \times 1.25 = 57.5A$$

$$\text{Total load} = 141.5A$$

From Table 310-16 a conductor with a minimum ampacity of 141.5A would be selected. A No. 1/0, THW copper conductor, with a 150A ampacity, would be capable of carrying 141.5A. The rules of Sec. 240-3, which require conductors to be protected against overcurrent in accordance with their ampacities would require a CB or fuses rated at a maximum of 150A. Although such an approach will help to simplify the task of assuring adequacy for the overcurrent protective device, it may also result in the selection of conductors that are larger than necessary.

In the example above, the service load (as determined in accordance with Art. 220) is 130A. In Table 310-16, it can be seen that a No. 1, THW copper conductor has an ampacity of 130A where there are not more than three conductors in a raceway and the ambient temperature is not over 86°F. According to Exception No. 4 to Sec. 240-3, the next standard rating of protective device—from Sec. 240-6—above the conductor's ampacity may be used. Therefore, it is permissible to protect the No. 1 THW copper conductor with a 150A protective device. Additionally, as can be seen above, the 150A overcurrent protective device would be properly rated for the combination of noncontinuous load plus 125 percent of the continuous load.

Location of Service Equipment

Secs. 230-2, 230-40, 230-70(a), 250-23(a), and 250-24(a). Providing additional capacity at an existing multi-building facility requires careful correlation of many Code rules.

Question: I would like your opinion regarding the application shown in Figure 1 and described below.

Three commercial buildings are located on one property, under single management and ownership. The electrical supply for each building originates at a pad-mount transformer that is centrally located among the buildings at a distance of approximately 400 ft. from each building. Each of the three buildings contains its own service equipment and service disconnect.

An addition was made to one of the buildings and increased capacity was required in that building. To supply the extra load, another 600A

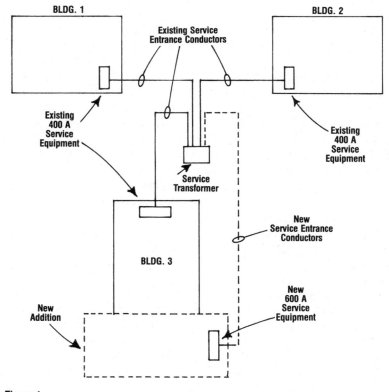

Figure 1

service was installed in the addition at the other end of the building from the existing 400A service equipment and disconnect.

The inspector was of the opinion that none of the Exceptions to Sec. 230-2—which permit more than one service to a building or structure under certain conditions—applied and that only one service would be permitted.

The electrical engineer for this project suggested that four fusible disconnects in NEMA 3R enclosures be installed on a plank at the outside transformer, which by Code definition would make those disconnects the "service equipment" and the conductors running to the buildings "feeder" conductors.

My questions are: Does the wording of NEC Sec. 230-70(a)—which covers location of the service disconnect means—recognize installation of the service equipment adjacent to the outside transformer which is located away from the building? If so, is the water pipe grounding electrode system in each building required to be connected to this service equipment?

Your comments on this application would be very much appreciated.

Answer: The answer to your first question is yes. The wording of NEC Sec. 230-70(a) is intended to, and does, recognize installation of the service equipment outside and remote from the building supplied. Although the wording of that section has been revised in a number of recent Code editions, commentary in available documentation and the present wording clearly indicate that a building's (or structure's) service equipment is *not* prohibited from being located outside and at some distance away. Therefore, service equipment (the four fusible switches) installed at the transformer would satisfy the requirements of Sec. 230-70(a). (If the unprotected service entrance conductors are brought into the building, then the wording of Sec. 230-70(a) requires the service disconnect to be located, either inside or outside, at the point where the service conductors enter the building.)

The answer to the second question about connection to the grounding electrode (water piping) system is also yes. However, considering the Code rules involved, it would seem as if there should be no problem with properly satisfying the Code-required grounding and bonding.

When the original installation was done, the Code would have required the service equipment in each building to be bonded and grounded as covered in Sec. 250-23(a). That is, at the service equipment, the grounded service conductor should have been connected to a grounding electrode system through a grounding electrode conductor; the grounded service conductor should have been bonded to the equipment ground bus within the service equipment; and the grounded service conductor should have at least one other connection to a grounding electrode located at the transformer or elsewhere outside. If all of those elements are in place—especially if the "additional" grounding electrode required outside is located *at* the transformer—then compliance with the Code-required grounding for the proposed new arrangement can be easily achieved.

First, the requirements of Sec. 250-23(a) would now apply to the four fusible disconnects. Therefore, the grounded service conductor must be brought into each disconnect and must be connected to the grounding electrode (additional "outside" ground rod previously required) through a grounding electrode conductor and the grounded service conductor must be bonded to the equipment ground bus within the fusible switches. It is worth noting that use of a single ground rod to supplement metal water piping, or as the grounding electrode system, is only permitted where a ground resistance measurement indicates that the made electrode has a resistance to earth that is less than 25 ohms. If not, then at least one additional rod must be provided (NEC Sec. 250-84).

Next, we look at the grounding arrangement required at each of the individual buildings.

As indicated in your question, the conductors extending from the loadside of the four fusible switches are no longer service conductors. According to Code definition these are now feeder conductors, so the rules of Sec. 250-24(a)—*not* Sec. 250-23(a), as was the case previously—are to be satisfied. The basic rule of Sec. 250-24(a) requires us to treat a feeder to another building in the same manner as we would a service. That is, the grounded conductor must be connected to the building's grounding electrode system through a grounding electrode conductor, and the grounded conductor must be connected with a properly sized bonding jumper to the equipment ground bus within the disconnect's enclosure. If this was already done, and it should have been, the rules of Sec. 250-24(a) are satisfied and the system is properly grounded and bonded.

Although the use of four fusible disconnects at the transformer would seem to satisfy all Code rules involved, there are still some questions.

This first one is, "Are disconnects required at the transformer for the conductors supplying buildings 1 and 2?"

Generally, it is not permissible to supply more than one set of service entrance conductors from a single drop or lateral. There is an exception that would recognize supplying multiple sets of service entrance conductors from a single drop or lateral, but only when all the service disconnects (two to six) are grouped at one location (NEC Sec. 230-40, Ex. No. 2). If there is only one service lateral, then disconnects would have to be provided for all four sets of conductors from the transformer to the building—the three existing and the new one—and be grouped at the transformer to satisfy the exception to the basic rule of Sec. 230-40. If, however, it is determined that there are actually multiple service laterals, then it would only be necessary to provide and group disconnects for the two "feeders" to BLDG 3. And the existing service laterals to BLDGs 1 and 2 may be left as is.

Based on the Code definition for a service lateral, as given in Article 100, it appears that there are actually multiple service laterals. A "service lateral" is defined as the "underground conductors...from a transformer" to the "first point of connection to the service-entrance conductors in a terminal box or meter, or other enclosure...." And, if there is no such "connection" of conductors between the transformer and the indoor service equipment, then the "point of connection" is to be taken as that point where the conductors extending from the secondary side of the utility transformer enter the building. Applying that definition to the original installation, it would seem that the conductors to each building are "service lateral conductors." Therefore, in the original installation there were actually three individual service laterals.

Now, back to the question of how many disconnects.

We now know each set of conductors supplied from the transformer secondary is considered to be an individual service lateral. And the wording of Sec. 230-40, Exception No. 2 would only require the multiple service disconnects supplied from any one service lateral to be grouped together. Therefore, the existing service laterals to BLDGs 1 and 2 may be left as is. And installation of two fusible disconnects, grouped at one location, would be considered the service equipment to BLDG 3. The conductors extending to BLDG 3 from the grouped fusible disconnects then become feeders (see definition of "feeder," Article 100) and there is no limit on the number of feeders that may be brought into a single building or structure. (Although not strictly required for feeder panels, because this installation is essentially a "two-service" arrangement, it would be a good idea to mark the main building disconnects in the same manner as would be required for the service disconnects where two separate services are located at different areas of a single building. A plaque or sign that indicates the location of the other main building disconnect and the service disconnects should be placed at each main building disconnect to alert operations, maintenance, and fire response personnel that opening a single main building disconnect does *not* completely isolate the building's electrical supply.)

There is one last concern. Generally, use of two sets of conductors—one to each disconnect—from the transformer secondary would be construed as a technical violation of the basic rule of Sec. 230-2—which permits only one service to supply a building or structure—unless the conductors are No. 1/0, or larger, and run underground to the disconnects (Sec. 230-2, Exception No. 7). If not, use of an individual set of conductors from the transformer secondary to each switch would technically be considered to be two services. If Exception No. 7 does *not* apply, a single set of conductors—properly sized according to Article 220 for the total load to be served—could be run from the transformer secondary into an enclosure where an individual set of service entrance conductors could be tapped to supply each disconnect.

Metal Conduit Protecting Service (or Feeder) Conductors on an Outdoor Pole Must Be Grounded *and* Arranged to Clear a Fault

Secs. 250-32, 250-51, 250-61, and 250-91. Much controversy is often generated by the subject of grounding metal conduit that is used as a protective sleeve on service conductors brought down a pole for an underground (lateral) service or feeder to a building or other structure. Article 250 has a number of very clear and widely understood rules that apply to this consideration. Figure 1 shows a typical application of the equipment under discussion here.

In the common layout for underground service from a pole line to a building, the conductors are run down the pole in nonmetallic raceway to a point at least 8 ft. above the ground—at which point metal con-

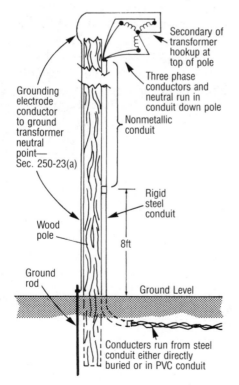

Secondary of transformer hookup at top of pole

Three phase conductors and neutral run in conduit down pole

Nonmetallic conduit

Grounding electrode conductor to ground transformer neutral point— Sec. 250-23(a)

Rigid steel conduit

Wood pole

8ft

Ground rod

Ground Level

Conducters run from steel conduit either directly buried or in PVC conduit

Figure 1

duit is commonly used for the run down and into the earth. Sec. 300-5(d) requires that the conductors be mechanically protected from below ground up to a height of at least 8 ft. aboveground by rigid metal conduit, intermediate metal conduit (IMC), or Schedule 80 rigid nonmetallic conduit. The underground run of conductors is then made directly buried or in nonmetallic conduit. In those cases where rigid metal conduit or IMC is used to protect the service conductors for the first 8 ft. up the pole, the questions arise: (1) Must the length of metal conduit be grounded? and, if so, (2) How must it be grounded?

NEC Sec. 250-32 says that "Metal enclosures for service conductors...shall be grounded," and that rule applies to service "raceways." As a result, the length of metal conduit that encloses service conductors on the pole must be grounded. There are no exceptions given in Sec. 250-32.

The basic elements of this problem must be clearly understood to properly satisfy the safety intents of the Code rules that apply:

1. A metal raceway enclosing service conductors run down a pole or on the outside of a building is always exposed to the possibility that one of the enclosed energized conductors might sustain an insulation failure that places the metal of the conductor in contact with the metal of the conduit (Figure 2).

2. When such contact between conductor and conduit takes place, a voltage (phase-to-neutral voltage) will be placed on the metal conduit. If there is not a low-impedance equipment grounding connection between the conduit and the neutral of the supply circuit, the fault con-

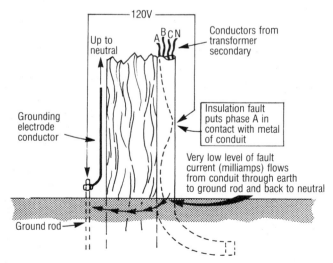

Up to
neutral

A B C N

Conductors from
transformer
secondary

Grounding
electrode
conductor

Insulation fault
puts phase A in
contact with metal
of conduit

Very low level of fault
current (milliamps) flows
from conduit through earth
to ground rod and back to neutral

Ground rod

Figure 2

nection will remain intact and the voltage on the conduit will pose a shock or electrocution hazard to persons who might touch the conduit and have another part of their body in contact with earth or some other metal that is in contact with earth.

3. The danger of that situation is not at all reduced by connecting the metal conduit to a driven ground rod at the pole. And because service conductors do not usually have overcurrent protection at their supply ends, there can never be clearing of the fault by opening of the fuse or circuit breaker. And, even if there were overcurrent protection in the circuit, the impedance of the earth would keep fault current far too low (milliamps or microamps) to operate any fuse or CB. (See last sentence of Sec. 250-51, which notes the inadequacy of current flow through the earth.)

4. The only way to clear dangerous potential (voltage) from the service conduit is to provide an equipment grounding connection between the conduit and the neutral conductor. If the metal conduit is connected to the neutral, any phase conductor coming in contact with the conduit will produce a phase-to-neutral fault that will carry enough current to burn open the point of fault and thereby remove the dangerous voltage (and shock hazard) [Figure 3].

5. This whole scenario emphasizes the need to provide effective return from metal enclosures to the grounded neutral of a grounded electrical system. It also pinpoints the fallacy that mere connection to

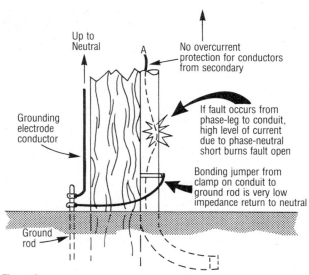

Figure 3

earth (such as to a grounding electrode) will provide for fault clearing and assure safety to persons. The last sentence of Sec. 250-51 and the last phrase of Sec. 250-91(c) prohibit use of the earth as a path of fault-current flow—that is, the earth must not be used as the "sole equipment grounding conductor."

Because NEC Sec. 250-61(a) permits use of a "grounded circuit conductor" (that is, a grounded neutral conductor of a circuit) to ground service conduits, such connection is the most effective way to satisfy Sec. 250-32, which requires grounding of metal service conduits. But, the question immediately arises, "How can that be accomplished?"

The first and most obvious method that could be used to bond metal service conduit to the neutral would be a bonding jumper from a lug on a clamp or fitting on the service conduit length—with the jumper connecting to a tap connector on the grounded neutral (Figure 3). With such a connection, the metal conduit would be grounded, as required by Sec. 250-32, without depending on earth as an equipment grounding conductor. Or, an equipment grounding conductor could be run from a connection on the conduit up to a neutral connection at the top of the pole at the supplying transformer.

For those cases where metal conduit is installed on a pole that does not have the supply transformer mounted at its top—where the service (or feeder) conductors are simply tapped from the overhead line conductors—the same clearing of a faulted phase leg can be accomplished by running an equipment grounding conductor from a clamp on the conduit to a tap into the grounded neutral conductor at the top of the pole. Or, a suitably protected bonding bushing could be used on the buried end of the conduit, with the underground neutral connected to the lug on the buried bonding bushing.

Grounding connections must be properly made to provide the safety of automatic clearing of a ground fault in a conduit on a pole. At left, the bonding jumper from the clamp on the conduit to the lug on the ground rod at left returns any fault current from the conduit to the transformer neutral. This is an actual hookup corresponding to the connections in the sketch of Figure 3. At right, connection is made from lugs on bonding locknuts (threaded onto the conduit ends before the couplings) to a driven ground rod, *but* if no connection is made from the conduits to a conductor that returns to the system neutral, then there is no provision for enough fault-current flow to burn open a fault contact between the metal conduit and an enclosed energized conductor. Connection to the ground rod is of no use and simply parallels the good earth contact the buried conduit ends already have.

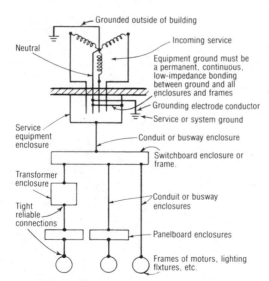

This shows the basic concept behind the phrase "equipment grounding system." For either a grounded or ungrounded electrical system, all metallic enclosures of conductors and other operating components must be interconnected and grounded at the service-system source.

Romex for Ranges and Dryers?

Secs. 250-60, 250-91(b), and 310-2(a). Where supplying 120/240V loads, the neutral must be insulated.

Question: Is it permissible to use a 10/2 Romex (Type NM) cable as the supply conductors for a 120/240VAC electric range? I think I've seen this done a couple of times, but I can't find where the Code talks about this. Can you help?

Answer: The use of 10/2 Type NM cable as the supply conductors for a 120/240VAC electric range is not permissible. This is covered in sections of the Code.

Because the bare conductor in the 10/2 Romex would also have to act as the neutral or "grounded" conductor in such a circuit to supply the 120VAC load, the fact the conductor is bare would constitute a violation of Sec. 310-2. In that section the NEC requires all grounded conductors to be insulated, unless specifically permitted elsewhere. No such "other" permission is given in the Code for Type NM cable.

Also, there is the concern for grounding. Generally, the Code prohibits the use of the "grounded conductor" for "grounding" equipment. Typically, it requires the use of one of the equipment grounding conductors described in Sec. 250-91(b) for equipment grounding. However, as covered in Sec. 250-60, the "grounded conductor" may be used as the equipment grounding means for "electric ranges and dryers" where certain conditions of use are satisfied.

One of the prerequisites for such applications given in part (c) of Sec. 250-60 calls for the "grounded conductor" to be insulated. And the bare or uninsulated conductor in a Type SE cable may also be used as the "grounded conductor," but only when the Type SE cable is run from the service panel directly to the range or dryer.

Therefore, because Sec. 250-60(c) does not recognize use of the bare conductor in Type NM for grounding a range or dryer, and because Sec. 310-2 generally requires all "grounded conductors" to be insulated, Type NM cable would not be permitted to supply a 120/240VAC electric range.

Bonding of Generator Output Circuits May Be Provided by the "Grounded" Conductor

Secs. 250-61, 250-79, and 250-91. **The grounded conductor (usually the neutral) of a circuit may be used to ground metal equipment enclosures and raceways on the supply side of the service disconnect or the supply side of the first disconnect from a separately derived generator output.**

A variety of tricky and often controversial details are involved in the design and installation of safe, effective, code-conforming grounding and bonding of the output circuit of emergency and standby generators. And a revision of the rule of Sec. 250-61 in the 1990 NEC applies to the very common use of flexible metal conduit enclosing the generator output circuit.

As shown in Figure 1, the output circuit from the generator is run in flexible metal conduit from the output terminal box of the generator up to the overhead junction box, where the conductors from the flex are connected to a feeder busway that runs to the generator disconnect and protection—located about 50 feet away, adjacent to the automatic transfer switch to which the generator circuit is connected. Flexible metal conduit is the common method for making this connection to such an overhead JB—from which the generator output circuit is run in either metal conduit, busway, or metalclad cable to the associated switching and control equipment. This flexible type of connection provides effective vibration isolation and, because of its flexibility, greatly facilitates ready connection.

Because of the size of the flex and the size of conductors used in parallel for the required ampacity of the generator output, the required equipment ground continuity between the generator output terminal box and the overhead junction box cannot be provided by the metal flex itself. Flexible metal conduit is not UL-listed as a general-purpose equipment grounding means, and NEC Sec. 250-91(b) permits a maximum 6-ft. length of metal flex to be used for equipment grounding *only* when the contained conductors are protected at 20A or less—which is not the case with generator output circuits.

Prior to the 1990 NEC, the only way to provide the required grounding continuity between the generator and the overhead junction box was to install properly-sized parallel equipment grounding conductors (which, in these short lengths are commonly called "bonding jumpers" or "bonding conductors")—either within or outside the flex lengths, as

Figure 1 Equipment grounding current path must be assured from the generator terminal box (lower arrow) to the overhead junction box (upper arrow), where the generator output circuit conductors connect to feeder busway that runs to the generator disconnect, protection and transfer switch.

covered by Sec. 250-79(f). Sizing of the parallel equipment bonding jumpers must satisfy Sec. 250-79(d)—which will be covered later. *But now,* Sec. 250-61(a)(3) will permit the parallel neutral conductors that are run within the flex lengths to provide for bonding the overhead junction box back to the generator terminal box and to the generator's bonded-neutral terminal point within the generator terminal box. This is accomplished by "bonding" (that is, "connecting") the neutral conductors within the overhead junction box to the metal of the box. Then any equipment ground-fault current originating within the electrical system being supplied by the generator—that is, when the generator and not the normal service is providing the power to the system—will return over the system's metal conduits and enclosures (or other equipment grounding conductors), coming back over the enclosure of the busway shown here, to the overhead junction box, from which the fault current will flow on the neutral conductors that are run through the flex lengths back to the generator bonded-neutral point. With such an arrangement, there is no need for bonding jumpers to be run with the flex lengths. Elimination of those bonding conductors removes significant material and labor costs, including the possibility that such bonding conductors would have dictated use of larger size of the flex.

Sec. 250-61(a)(3) permits the neutral (grounded) conductor of a circuit to be used to ground metal equipment enclosures and raceways that are "on the supply side" of the disconnect or overcurrent protection of a "separately derived system." When a generator is used with a 4-pole transfer switch—one that has a switch pole for the neutral conductor, so there is not a connection between the generator neutral and the neutral of the normal supply—and that is almost always the arrangement used today—the generator is then a "separately derived system" by Code definition. As such, the generator output circuit is eligible for the permission given in Sec. 250-61(a)(3) to use the neutral as an equipment grounding conductor.

Properly Conducting Ground-Resistance Measurements Require Special Techniques and Tools

Sec. 250-84. NEC rule requires measuring "ground resistance" of the grounding electrode system where a single driven rod is used.

Grounding for safety has been a concern with us since Ben Franklin invented the lightning rod. However, when it comes to performing ground resistance measurements, many designers and installers are not exactly sure what is involved. To a certain extent, measuring of the grounding electrode resistance is something of a mystery. And, generally, there has been little guidance provided. However, the advent of modern technology has required us to attach new importance to proper installation of the safety grounding system.

All too often, the aura of "mystery" surrounding these measurements causes those who should be concerned with such measurements to shy away from performing the task. In light of all the conflicting data about ground resistance measurements, this shyness is understandable. Indeed, this conflicting information is a result of there being no hard-and-fast rule that applies to all circumstances and conditions. But, once one understands the true nature of the task, it is then possible to apply this knowledge to the prevailing conditions and obtain reliable results.

What is "ground resistance?"

The term "ground resistance" (often referred to as "ground rod resistance") is the value, measured in ohms, of the resistance between a grounding electrode and a reference ground rod located at a remote distance from the grounding electrode. That is, ground resistance is the ohmic value of resistance between the driven grounding electrode and a remote "test rod" as measured through the earth. A well-accepted explanation of the earth as a conductor—which it is during the ground resistance measurement—is to consider the earth (ground) surrounding the electrode as being made up of a series of conductive shells (as shown in Figure 1).

As can be seen, the earth shell closest to the ground rod has the smallest surface area, and, as a result, offers the greatest resistance to current flow. Succeeding shells have a greater surface area and offer less resistance. At some remote distance from the grounding electrode, the surface area of the concentric earth shells is so great that no significant additional resistance is added to that of the earth.

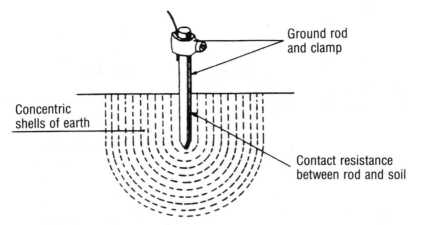

Figure 1 Resistive components of an earth electrode.

From this we see that the "ground resistance" is the contact resistance of the rod—which is negligible where a clean, grease- and paint-free rod is used—and that of the earth or "ground" as measured at some distance from the grounding electrode. Therefore, the "ground resistance" for any electrode will depend heavily on the actual resistance of the soil (the soil resistivity) surrounding the electrode.

Soil resistivity indicates the resistance of the soil based on a unit volume of the particular soil. That is, it indicates the resistance of a cubic measure (cubic-inch; cubic-centimeter; cubic-foot; etc.) of soil, as measured on opposite faces of the soil cube. Soil resistivity is expressed in ohm-inches, ohm-centimeters, etc., and represented by the Greek letter ρ (rho).

When designing an extensive grounding system to assure a certain minimum "ground resistance," knowledge of the soil resistivity (its ability to conduct an electric current) becomes an extremely important requirement. This may be accomplished through the use of a specific test and manipulation of a specific formula (Figure 2), or the general information presented in Table 1 of Figure 3 may prove to be adequate.

As might be expected, in addition to the "type" of soil, the moisture content as well as the ambient temperature will affect the soil resistivity. Tables 2 and 3 in Figure 3 show how these variables can affect the actual soil resistivity. Obviously, use of the test method described in Figure 2 will provide the most precise indication of soil resistivity at any specific location for use in designing a grounding electrode system that will result in the desired "ground resistance."

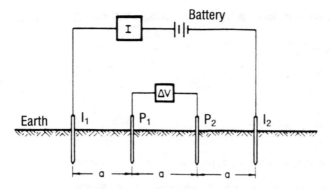

The resistance (R) is determined from Ohm's law, as follows:

$$R = \frac{\Delta V}{I}$$

And, the soil resistivity is calculated from the following:

$$\rho = 2\pi a R$$

where: a = spacing of electrodes in feet
        R = measured resistance as
            determined by Ohm's law:

Figure 2   Soil resistivity is determined using the "Wenner array" (as shown above). The array consists of four electrodes, placed in a straight line, at equal distances from each other. The Wenner array measures the potential difference between the two middle or inside electrodes (P_1 and P_2), which is caused by passing a current from one outside electrode to the other (I_1 and I_2). Although a DC current may be used, an AC source is preferable because the polarity of the DC source must be reversed and the measurement must be repeated to assure an accurate reading. It should be noted that the formula shown here is only useful when the distance between the electrodes (a) is equal to 20 times the driven depth of the electrodes. When the spacing distance is closer, a more complex formula (not shown) must be applied. Therefore, assure that the distance a is equal to or greater than 20 times the driven depth of the electrodes.

Measuring ground resistance

Over the years a number of erroneous "methods" have been used by many electrical designers and installers. For instance, some use a fuse to evaluate the effectiveness of the grounding electrode. Typically, they would place a 5A fuse between one of the ungrounded (hot) conductors and the ground rod. If the fuse blew, everything was fine. Another commonly used (and equally inadequate) method for testing the grounding electrode employed a 75W incandescent bulb. The bulb was connected between the grounding electrode and an ungrounded conductor, and if it appeared to glow "normally," the grounding electrode arrangement was considered to be satisfactory.

Soil	Resistivity, Ohm-cm (Range)		
Surface soils, loam, etc...............	100	—	5,000
Clay..................................	200	—	10,000
Sand and gravel......................	5,000	—	100,000
Surface limestone....................	10,000	—	1,000,000
Limestones...........................	500	—	400,000
Shales................................	500	—	10,000
Sandstone............................	2,000	—	200,000
Granites, basalts, etc.................		100,000	
Decomposed gneisses.................	5,000	—	50,000
Slates, etc...........................	1,000	—	10,000

Table I

Moisture Content, % By Weight	Resistivity, Ohm-cm	
	Top Soil	Sandy Loam
0	$1,000 \times 10^6$	$1,000 \times 10^6$
2.5	250,000	150,000
5	165,000	43,000
10	53,000	18,500
15	19,000	10,500
20	12,000	6,300
30	6,400	4,200

Table II

Temperature		Resistivity, Ohm-cm
C	F	
20	68	7,200
10	50	9,900
0	32 (water)	13,800
0	32 (ice)	30,000
−5	23	79,000
−15	5	330,000

Table III

Figure 3 As shown in Tables I, II, and III, the type of soil, its moisture content, and its temperature all influence the soil resistivity.

In both of the above-mentioned incorrect and unreliable "methods," the grounding electrode conductor was connected to the neutral bus or block in the service equipment. Therefore, the grounding electrode conductor provided a return path for current flow when the ungrounded conductor was connected through the fuse or bulb to the grounding electrode, and, as a result, such "tests" always indicated that everything was okay.

Another "method" that appears to be rather popular—although I wish it weren't—is the use of a megohmmeter (the so-called "megger"). This piece of equipment is designed to perform insulation-resistance measurements, which are generally in the millions-of-ohms range, hence the name, megohmmeter ("meg," the abbreviation for *mega,* which means *million;* "ohm," the unit of measure for resistance; and "meter," a measuring device). Generally, the megohmmeter is hooked up to the electrode and the nearest water pipe, or, in some instances, even to a screwdriver pushed into the ground.

Some who have used this method will say, "The meter showed zero impedance. Are you saying the meter's wrong?" Not at all. However, it must be remembered that the megohmmeter is designed to read resistances in the millions-of-ohms range. A 500V megohmmeter typically has an upper range of 100,000,000 ohms. That means the scale is divided into 100,000,000 parts. Just imagine trying to locate 25 parts (25 ohms) on a 3-inch scale divided into 100,000,000 parts. Obviously, the meter will appear to show zero resistance every time.

If those methods are wrong, how does one go about properly measuring "ground resistance"?

Although there are some relatively exotic means for measuring ground resistance—such as those developed by the Mine Safety and Health Administration—those methods are not readily adaptable for general use. The traditional method for properly determining ground resistance is the "three-point" or "fall-of-potential" method. Because the fall-of-potential method is the most commonly used, the following discussion covers the equipment and technique for properly executing a ground-resistance measurement using the fall-of-potential method.

The instrument

There are a number of factors that can affect the accuracy of a ground-resistance measurement. As a result, the measuring device should be designed in such a way as to prevent those factors from impairing or diminishing the accuracy of the measurement. Generally, the factors of concern are:

1. *Stray ground currents from other systems.*   These can come from numerous sources—utility grounds, fault currents, electric train rails, communication circuits, and lightning discharges. Generally, stray currents are at 60Hz and harmonics of 60Hz. Therefore, the test equipment must be designed to reject those frequencies.

2. *Polarization.*   When current is passed between two electrodes in an electrolyte, gas bubbles form on the surface of the electrodes, which increases the apparent resistance. Because the earth is essentially a "semi-electrolyte," where DC current is used to perform the ground-resistance measurement, a similar phenomenon occurs. If a DC current is used, always reverse the polarity of the DC current source and average the readings to minimize the affect of po-

larization. Use of equipment that generates AC current eliminates concern for polarization.

3. *Galvanic action.* This phenomenon occurs where two dissimilar metals are placed into an electrolyte. The test rods are usually galvanized steel. If the grounding electrode under test is aluminum, galvanic action can affect the measurement. Again, this is only a concern where the current source is DC.

4. *Nature of the soil.* That is, soil resistivity, moisture content, and temperature.

Therefore, the test instrument should be capable of generating an AC current at a frequency other than 60Hz. It should contain filtering to reduce the affects of stray currents and block DC currents. The test instrument should also contain protection against unexpected fault currents and have adequate sensitivity to operate at relatively low signal levels.

Note: Although the diagram in Figure 4 appears to show two different test instruments, today's commercially available ground-resistance measurement instruments combine both the current source and the voltmeter in a single instrument. And such instruments will give a calibrated readout in ohms, which eliminates the need for any calculations to determine the actual ground resistance.

Setup and measurement

Before discussing the specifics regarding setup and testing, there are some safety precautions that must be observed.

Prior to connecting any test leads to the grounding electrode under test, assure that the grounding electrode conductor is not connected to the grounding electrode. Because most resistance testing is conducted before the system is energized, this generally will not present a problem. However, for an existing energized system, the main breaker must be opened and the grounding electrode conductor should be checked for current flow prior to disconnecting the grounding electrode connection. If there is current flow on the grounding electrode conductor, the equipment producing the current must be located and deenergized. Failure to do so could present additional shock hazards and will result in an inaccurate measurement. Remember, if the grounding electrode conductor is not disconnected, you will be reading the ground resistance of the entire system (including any other electrodes—such as metal water piping, building steel, rebars), not just the grounding electrode that is supposed to be tested.

Once the test leads have been connected to the electrode, the operator should wear high-voltage rubber safety gloves and, to the maxi-

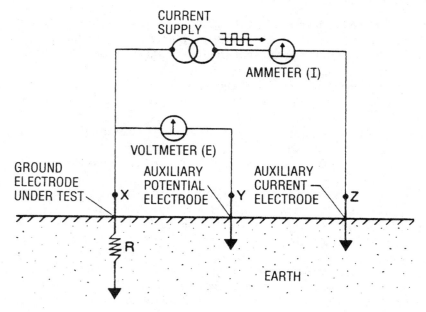

CURRENT
SUPPLY

AMMETER (I)

VOLTMETER (E)

GROUND
ELECTRODE
UNDER TEST

AUXILIARY
POTENTIAL
ELECTRODE

AUXILIARY
CURRENT
ELECTRODE

X   Y   Z

R

EARTH

Figure 4 The fall-of-potential method for ground-resistance measurement essentially measures the difference of potential between the electrode under test (at point X) and the "potential" or "voltage" electrode (point Y). As shown above, an AC current (could be DC, although AC is preferred) is passed from the "current" electrode (at point Z) to the electrode under test (point X). The voltmeter measures the voltage between the test electrode and the "potential" or "voltage" electrode. Then, the resistance (R) is determined from Ohm's law: $R = E/I$, where E is the voltage measured between points X and Y, and I is the current passed between X and Z (as shown by the ammeter reading).

mum extent possible, refrain from handling the test instrument. The best thing is to place the instrument on a table or chair and leave it there until the test is complete. This is an especially important concern because if a major fault occurs in the vicinity of the test—such as if a utility power line faulted—there could be a lethal potential between the rods and earth. Although many choose to ignore these simple safety rules, they do so at their own peril as the hazards are ever present and can be fatal.

Once the grounding electrode has been isolated from all other portions of the electrical system, the test is set up as shown in Figure 4. As can be seen, in addition to the rod under test, two additional rods must be driven: the "current" electrode and the "potential" or "voltage" electrode.

Placement of the test electrodes for making a ground-resistance measurement is all important. The concern here is related to the fact that electric currents through the earth flow in paths very similar to that of the lines of force emanating from a magnet. Recall the patterns of iron filings sprinkled over a piece of paper covering the magnet. The lines of force emanate at right angles from one pole and transverse and arc entering the opposite pole at right angles. Also recall that the maximum concentration of force lines is in the direct path from one pole to the other. From this simple analogy, it can be seen that to obtain the best results—maximum current flow between rods to assure maximum sensitivity—the electrodes should be, as nearly as possible, in a straight line.

In addition to assuring that the three rods are in as straight a line as possible, accuracy of the test depends on placement of both the "current" and "voltage" electrode at the correct distances from the electrode under test.

Placement of the "current" electrode must be such that the mutual resistance or coupling between the "current" electrode and the rod under test is minimized. This task would be greatly simplified if the soil resistivity were completely uniform and there were no underground metal piping systems in the vicinity. Unfortunately, this is generally not the case. And, to a certain extent, determining the optimum distance will be a matter of trial and error, although there is a rule-of-thumb based on electrode under test. That is, use a distance equal to 10 times the driven length of the electrode as a starting point [i.e., for a completely buried 8-ft. rod, install the "current" electrode approximately 80 ft. away from the electrode under test; for a partially buried 10-ft. rod (8-ft. minimum), approximately 80 ft. away; and for a completely buried 10-ft. rod, approximately 100 ft. away, etc.].

Location of the "voltage" electrode is most commonly based on a percentage of the distance between the "current" electrode and the electrode under test. Known as the "62 percent method," it calls for installation of the "voltage" electrode at a distance equal to 62 percent of the distance between the electrode under test and the "current" electrode. If the "current" electrode is 100 ft. away from the electrode under test, then the "voltage" electrode should be driven at a distance of 62 ft. from the electrode under test.

As previously indicated, after the grounding electrode under test has been completely isolated from the rest of the electrical system, the test instrument leads are connected. The X terminal of the test instrument is connected to the grounding electrode under test. The Z terminal of the instrument is connected to the "current" electrode, and the Y terminal is connected to the "voltage" electrode.

At this point, a reading is taken. Next, two additional readings are taken with the "voltage" electrode located 12 ft. fore and aft of the 62-ft. point, that is, the "voltage" electrode is moved and measurements are made at 50 and 74 feet from the electrode under test (the "current" electrode remains in place). Those readings should be within 10 percent of the value indicated at the 62-ft. distance, and the 50-ft. test value—the measurement where the voltage electrode is closest to the rod under test—should be the lowest value. Then, the three readings are averaged and the average represents the ground resistance of the electrode under test.

Any variation greater than those parameters indicates improper placement of the "current" electrode. Therefore, the "current" electrode must be relocated at a greater distance from the electrode under test and the measurement must be repeated. Or, because underground metallic piping can significantly affect the accuracy of the measurement, it may be necessary to move the "current" electrode to a different position on the periphery of an arc that has the electrode under test as its center (Figure 5). Obviously, the "voltage" electrode would also have to be relocated in order to achieve the same arrangement described above and the measurements would have to be repeated.

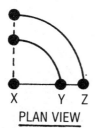

X   Y   Z

PLAN VIEW

Figure 5 In some cases, the presence of underground metallic piping systems may adversely affect the test measurement. Relief from "interference" by such underground systems can usually be achieved by moving the "current" and "voltage" electrodes as shown above.

Another method, which yields a higher degree of accuracy, involves plotting the measured resistance at different distances from the electrode under test. That is, the "current" electrode is located as described before (at a distance equal to 10 times the length of the electrode under test that is in contact with the earth) and the "voltage" electrode is placed at a distance of 5 ft. from the electrode under test. A measurement is made and the value recorded. The "voltage" electrode is then moved another 5 ft. closer to the "current" electrode and another measurement is made. This procedure is continued until the

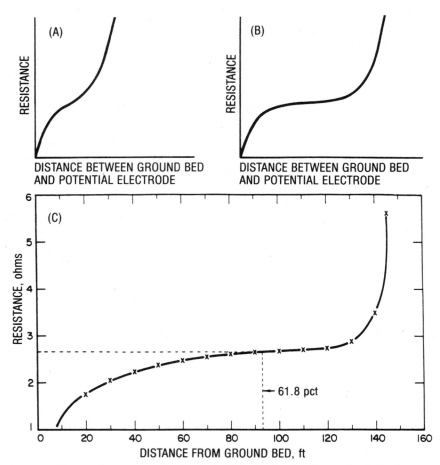

Figure 6 The graph shown at the top left (A) indicates the "current" electrode is too close to the electrode under test. The graph at top right (B) is the expected result where the "current" electrode is located at a proper distance from the electrode under test. The graph at bottom (C) is an actual plot of a very low resistance ground bed constructed using multiple rods, which is often required in a mining or quarrying operation to assure maximum safety.

"voltage" electrode is within 5 ft. of the "current" electrode. The measured values are then plotted as "Resistance vs. Distance" (Figure 6). Note that the midpoint of the resulting curve is a flat line or plateau where the resistance is fairly constant. That indicates the "ground resistance" of the grounding electrode.

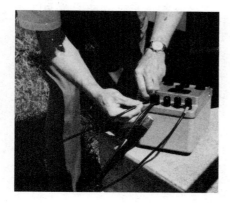

This represents the basics regarding the proper technique to use for correctly determining "ground resistance." Although mastering of all the nuances and idiosyncracies related to ground-resistance measurements will require time and experience, compliance with the procedures and recommendations given here should prove adequate for many of the more routine applications. If you experience problems with any specific installation, the manufacturer or salesperson representing the ground-resistance test equipment will generally be happy to provide guidance about what may be the cause and suggest corrective action. However, there is no substitute for personal experience. The best way to become proficient is to get out and use the equipment and procedures.

The NEC and "ground resistance" measurements

As most are aware, the National Electical Code (NEC) contains very detailed and specific requirements for the equipment, methods, and arrangement of the individual elements that makeup the safety grounding system. Basically stated, the Code requires interconnection of the "grounded conductor" to all noncurrent-carrying metallic parts of the system through the "equipment grounding conductor" and, to the "grounding electrode system" through the "grounding electrode conductor"

As covered in Part H of Article 250, the "grounding electrode system" may take many forms. It may be a combination of a metallic water piping and building steel and rebars in the concrete footing or foundation of a building. In the absence of building steel and rebars in the footing or foundation, we are required to provide a "made electrode" to supplement the water piping. Where none of those "electrodes" are available, we would be required to use a "made electrode" (typically-ing a driven ground rod) to provide for a connection to earth or ground.

The primary purpose for requiring connection to earth through the grounding electrode system is to allow for lighting discharge, although such connection also helps stabilize system voltage and helps assure that there is no difference of potential between the noncurrent-carrying metal of the electrical distribution system and building steel or an interior metal water piping system.

Where a single driven ground rod is used—either to supplement the metal water piping system or because no other electrode is available—the NEC requires that the "ground resistance" be not greater than 25 ohms. If it is, then one additional made electrode must be installed at least 6 ft away from the first and connected using not less than a No. 6 copper or No. 4 aluminum conductor (Secs. 250-84 and 250-94, Ex No. 1a).

Although the wording used in Sec. 250-84 is commonly interpreted to recognize the use of two ground rods *without* making *any* measurement, it must be remembered that the NEC is a bare minimum safety standard; it is *not* a design guide. For many of today's installations, the ground resistance must be limited to certain specific values. And the only way to accurately determine the ground resistance in any specific application is to measure it. The discussion presented here is intended to illuminate the overall task and specific procedures necessary to assure reliable and repeatable results where a sure and confident knowledge of the actual "ground resistance" is desired or required by the designer and/or installer.

Minimum (?) Size of Equipment Grounding Conductor

Sec. 250-95 and Table 250-95. For many years now, there has been considerable discussion among electrical designers and installers regarding proper sizing for an equipment grounding conductor as covered by the NEC in Sec. 250-95 and Table 250-95. That discussion has centered around the meaning of the word "minimum" found in the table heading. But what is the "minimum" if high fault currents are available?

Over the years, most designers and installers have considered the minimum size of equipment grounding conductor given in Table 250-95 to be adequate regardless of the short-circuit current available at that point in the distribution system. That is, whether there were 5,000 or 50,000A of available fault current, generally the same size of equipment grounding conductor would be selected. While this may not have been an issue years ago—very few installations had available short-circuit currents than were greater than 5,000 or 10,000A—today, with much larger distribution systems and lower-impedance transformers (some below 1 percent) it is no longer unusual to see available fault currents in excess of 100,000, and in some cases, over 200,000A. It is those installations where high levels of short-circuit current are available that pose the real problem and require additional consideration.

The concern for providing protection for the equipment grounding conductor was recognized a number of years ago, and was explained

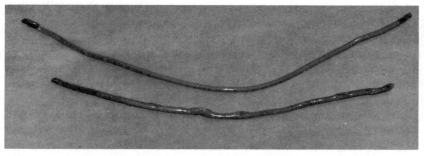

Here are two No. 12 AWG, THHN, copper conductors after being exposed to 40,000A. The one No. 12 (at top) was protected in accordance with ICEA limits by a current-limiting overcurrent protective device. The other was not. As can be seen, in the first case, the conductor was not damaged because the current-limiting protective device opened the circuit before insulation temperature limits were exceeded. In the other conductor (at bottom), the operating characteristics of the overcurrent protective device resulted in a value of current flow that severely damaged the THHN (90°C-rated) insulation.

by Eustace Soares in his now famous work entitled, "Grounding Electrical Distribution Systems for Safety." In that book, Mr. Soares indicates that the "validity" of a conductor as a fault return path is maintained only where loaded not in excess of a specific current value for a specified amount of time based on the conductor's cross-sectional circular mils. That is, for a copper conductor, its integrity or "validity" is maintained where the amount of current for each 30 circular mils of cross-sectional area is not greater than 1A and does not persist for more than 5 seconds. That "validity" is related to the amount of energy that would cause the copper conductor to become loosened at its point of attachment after the copper cools to ambient. That validity rating is based on raising the temperature of the copper conductor from 75°C to 250°C.

Another recognized method looks at the amount of energy that will produce damage to the conductor's insulating material. The method to which I refer is covered in the International Cable Engineers Association publication P-32–382. That method calculates the energy required to raise the conductor's temperature from 75°C to 150°C, which will cause damage to the conductor's insulation. And a third method promoted by Onderdonk calculates the amount of energy necessary to cause the temperature of the conductor material to rise from 75°C to 1083°C, which will cause the copper to melt. That is, the Onderdonk method calculates the conductor melting point. Values of short-circuit current and withstand ratings calculated in accordance with each of those recognized methods are shown in Table 1.

But, does the NEC *require* that equipment grounding conductors be evaluated with regard to the available fault currents? In answer to that question, consider the following.

NEC Sec. 240-1, which indicates the "Scope" of Art. 240, is followed by a Fine Print Note (FPN) that says, "overcurrent protection is provided to open the circuit if the current reaches a value that will cause an excessive or dangerous temperature in conductors or conductor insulation." Although FPNs do not present a rule, they are included to provide "explanatory" information regarding certain Code requirements [see NEC Sec. 110-1]. The FPN following Sec. 240-1 indicates the overall concept behind overcurrent protection. Although an equipment grounding conductor is not normally carrying current, under fault conditions, it becomes part of "the circuit" and really should not be damaged any more than the phase or neutral conductors of a given circuit. But that is not the only Code reference that applies here.

The NEC in Sec. 250-51 requires that each and every equipment grounding conductor be capable of carrying any fault currents likely to be imposed upon it. And compliance with that requirement would necessitate some evaluation of the selected size of grounding conduc-

TABLE 1

COMPARISON OF EQUIPMENT GROUNDING CONDUCTOR SHORT CIRCUIT WITHSTAND RATINGS.

CONDUCTOR SIZE	5 SEC RATING (AMPS)			I²t RATING X10⁶ (AMPERE Squared Seconds)		
	ICEA P32-382 INSULATION DAMAGE 150 C	SOARES 1 AMP/30 cm VALIDITY 250 C	ONDERDONK MELTING POINT 1083 C	ICEA P32-382 INSULATION DAMAGE 150 C	SOARES 1 AMP/30 cm VALIDITY 250 C	ONDERDONK MELTING POINT 1083 C
14	97	137	253	.047	.094	.320
12	155	218	401	.120	.238	.804
10	246	346	638	.303	.599	2.03
8	391	550	1015	.764	1.51	5.15
6	621	875	1613	1.93	3.83	13.0
4	988	1391	2565	4.88	9.67	32.9
3	1246	1754	3234	7.76	15.4	52.3
2	1571	2212	4078	12.3	24.5	83.1
1	1981	2790	5144	19.6	38.9	132.
1/0	2500	3520	6490	31.2	61.9	210.
2/0	3150	4437	8180	49.6	98.4	331.
3/0	3972	5593	10313	78.9	156.	532.
4/0	5009	7053	13005	125.	248.	845.
250	5918	8333	15365	175.	347.	1180.
300	7101	10000	18438	252.	500.	1700.
350	8285	11667	21511	343.	680.	2314.
400	9468	13333	24584	448.	889.	3022.
500	11835	16667	30730	700.	1389.	4721.
600	14202	20000	36876	1008.	2000.	6799.
700	16569	23333	43022	1372.	2722.	9254.
750	17753	25000	46095	1576.	3125.	10623.
800	18936	26667	49168	1793.	3556.	12087.
900	21303	30000	55314	2269.	4500.	15298.
1000	23670	33333	61460	2801.	5555.	18867.

Table 1 Values of current (at left) and the let-through energy (I²t) withstand ratings for copper conductors as determined in accordance with three industry-recognized methods. Designers and installers should be aware that equipment grounding conductors selected in accordance with Sec. 250-95 and Table 250-95 of the NEC may *not* be adequate where large amounts of short-circuit current are available. In such applications, safety concerns for personnel and equipment dictate that the size of equipment grounding conductor be increased above the Code-recognized "minimum."

tor with respect to the available fault current. That is, in addition to assuring compliance with Sec. 250-95 on the minimum sizing, the conductor selected in accordance with that Code section must have sufficient cross-sectional area "to conduct safely" the fault current it will be required to carry.

Although the phrase "to conduct safely" is not very clear, use of one of the three preceding methods—either the Soares "validity" of connection method, the ICEA "insulation damage" method, or the Onderdonk "melt-point" method—should serve to satisfy the wording of Sec. 250-51. Of course, if the authority having jurisdiction prefers one method over the other, then the method preferred by the local inspector should be used.

Which is the more desirable and realistic approach? That depends on the application. For example, in most applications it would seem that Soares' "validity" of connection method is adequate. But, for isolated ground applications, the ICEA "insulation damage" method might be more appropriate to assure the desired isolation of the ground-return path even after a short-circuit. The "melt-point" method would appear to be the least desirable, but could be construed as satisfying the wording of Sec. 250-51(2). Regardless of which method is used, certainly some evaluation of the grounding conductor's fault current-carrying capability must be performed where there are large values of available short-circuit current to assure compliance with the NEC and to assure continuity of the ground-return path, which provides for automatic clearing of a faulted circuit.

How does one analyze the equipment grounding conductor for fault-carrying capability?

First, you will need to refer to the short-circuit analysis—which should have been performed to determine the minimum interrupting rating of the selected protective devices—and find the available fault current at that point in the system where the circuit originates. Next, using the protective device manufacturer's operating characteristics data, determine the amount of short-circuit current and the amount of let-through energy (I^2t) that the grounding conductor will be exposed to during a bolted fault. Then, refer to the data given in Table 1 to verify that the equipment grounding conductor, selected in accordance with Sec. 250-95 and Table 250-95, will be capable of carrying that value of current and withstanding the let-through energy (I^2t). If not, then a larger-sized conductor—that is, one capable of sustaining that value of current and withstanding the I^2t value—must be selected.

Although the problem of excessive fault current for the size of grounding conductor is more of a concern for feeders, branch circuits that originate in close proximity to the service equipment—such as at motor control centers located near the service equipment—should also

Careful consideration should be given to sizing of the equipment grounding conductor where large amounts of short-circuit current are available. The equipment grounding conductors selected for the lighting branch-circuits supplied above (arrow), which are fed from a panel in close proximity to the service equipment, should be evaluated to assure the selected grounding conductors will be capable of withstanding the value of fault current they are likely to carry.

be evaluated. Only through such an evaluation is it possible to assure compliance with the NEC and assure that the equipment grounding conductor will, in fact, be capable of facilitating operation of the circuit overcurrent protective device and provide automatic clearing of faulted conductors.

Beware When Using Flex and Liquidtight Flex "Where Flexibility Is Required"

Secs. 250-91, 350-4, 350-5, 351-8, and 351-9. Use of flex and liquidtight flex as the ground return path must satisfy NEC rules and UL limitations.

In all applications of flexible conduits, it is essential to carefully observe the need to provide equipment grounding for the equipment (motor, light, fixture, transformer, etc.) being supplied. There have been many shock and electrocution accidents due to failure to provide a solid, reliable, low-impedance return path for ground faults occurring in metal equipment enclosures fed by one of the flexible type raceways. We have encountered this problem in court litigation where electrical designers and/or installers have been charged with failing to provide adequate grounding in accordance with the rules of the NE Code and "prevailing safe practices."

Although the NE Code recognizes use of rigid metal conduit, intermediate metal conduit (IMC), and EMT as an equipment grounding

EQUIPMENT GROUNDING through this short length of flex is provided by an internal equipment grounding conductor from the electric solenoid valve (bottom, left) up to the wireway. Because the flex is not clamped, the rules of Sec. 351-8, Exception No. 2 and Sec. 351-9, Exception No. 1 combine to make use of the equipment ground wire (which could have been external) mandatory.

LIQUIDTIGHT FLEX from each starter to a motor is a violation of Sec. 351-8 because unclamped flex length over 3 ft. is not permitted for anything other than a lighting fixture whip, as noted in Exception No. 3. If there were one clamp in the middle of each run—so that the unclamped length of each run is not over 3 ft.—it would satisfy Exception No. 2, but an equipment grounding conductor would be required, either inside or outside the flex.

conductor for electrical systems [Sec. 250-91(b)], there is not general recognition of spiral-wound (interlocking) metal of flex or liquidtight flex as an equipment grounding conductor. The problem arises from the fact that current flow over such metal flex tends to follow the convolutions of the interlocked metal and such current is flowing, in effect, through a coil that has high inductive reactance, opposing current flow. The resulting high impedance of the current path makes the metal flex inadequate as an equipment grounding conductor. This is the same problem that makes it necessary to use the No. 16 aluminum bonding strip under the armor of BX cable; and to use a grounding conductor in interlocked type MC cable.

The whole subject of "grounding and bonding with flexible conduits" is a fairly complex topic as a result of the rules in the National Electrical Code. Those rules are covered in certain specific sections of the NE Code as follows:

For Flexible Metal Conduit   Sec. 350-5 covers "Grounding," and must be coordinated with Sec. 350-4 on "Supports" (clamps) required on the flex.

For Liquidtight Flexible Metal Conduit   Sec. 351-9 covers "Grounding," and must be coordinated with Sec. 351-8 on "Supports" (clamps) required on the conduit.

For Liquidtight Flexible Nonmetallic Conduit Sec. 351-27 covers "Equipment Grounding." An equipment grounding conductor must be used with nonmetallic conduit to assure safety grounding of the enclosure fed. Nonmetallic liquidtight is not permitted to be used in lengths over 6 feet and there is no requirement for clamping.

Equipment grounding for flexible metal conduit and liquidtight flexible metal conduit is also regulated by Exceptions No. 1 and No. 2 of Sec. 250-91(b). Those rules basically repeat the rules of Sec. 350-5 and Sec. 351-9.

Standard metal flex

Standard flexible metal conduit (also known as "Greenfield") is not listed by UL as suitable for grounding in itself. However, Sec. 350-5 of the NE Code as well as Exception No. 1, part (b) of Sec. 250-91 permits flex to be used without any supplemental grounding conductor when any length of flex in a ground return path is not over 6 ft. and the conductors contained in the flex are protected by over-current devices rated not over 20A (Figure 1). Use of standard flex with either internal or external bonding must be as follows:

1. When conductors within a length of flex up to 6 ft. are protected at more than 20A, equipment grounding may not be provided by the flex, but a separate conductor must be used for grounding. If a length of flex is short enough to permit a bonding jumper not over 6-ft. long

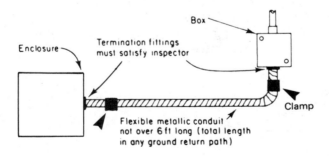

Flex not over 6 ft long is suitable as a grounding means (without a separate ground wire) if the conductors in it are protected by OC devices rated not more than 20 amps.

Figure 1 Under these conditions, with a clamp (arrow) within 12 in. of flex end, flexible metal conduit may be used without an equipment grounding conductor. Use of the flex without the clamps shown would be permitted by Exception No. 2 of Sec. 350-4—*but only* if "flexibility is necessary" at terminals. And if the clamps are omitted, an equipment grounding conductor (or bonding conductor) must be used either inside or outside the flex itself.

to be run between external grounding-type connectors at the flex ends, while keeping the jumper *along* the flex, such an external jumper may be used where equipment grounding is required—as for a short length of flex with circuit conductors in it protected at more than 20A. Of course, such short lengths of flex may also be "bonded" by a bonding jumper inside the flex, instead of external.

Figure 2 shows how an external binding jumper must be used with standard flexible metallic conduit (so-called Greenfield). If the length of the flex is not over 6 ft. and the conductors run within the flex are protected at more than 20A, a bonding jumper *must* be used either inside or outside the flex. An outside jumper must comply as shown. For a length of flex *not* over 6 ft., containing conductors that are protected at *not* more than 20A and used with conduit termination fittings that are approved for grounding, a bonding jumper is not required. Flex in any length over 6 ft. is not suitable as an equipment grounding conductor and an equipment grounding conductor must be used *within* the flex to ground metal enclosures fed by flex.

Exemption from the need for an equipment grounding conductor applies only to flex where there is not over 6 ft. of "length in any ground return path." That means that from any branch-circuit load device—lighting fixture, motor, etc.—all the way back to the service ground, the total permitted length of flex without a ground wire is 6 ft. In the total circuit run from the service to any outlet, there could be one 6-ft. length of flex, or two 3 ft. lengths, or three 2-ft. lengths, or a 4-ft. and a 2-ft. length—where the flex lengths are in series as equipment ground return paths. In any circuit run—feeder to subfeeder to branch

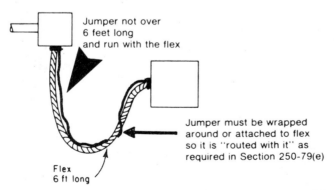

Jumper not over
6 feet long
and run with the flex

Jumper must be wrapped
around or attached to flex
so it is "routed with it" as
required in Section 250-79(e)

Flex
6 ft long

Figure 2 If the conductors within a flex run are protected by fuses or CB rated *over* 20A, a bonding jumper (equipment grounding conductor) must be used even if the length of the flex is *not* over 6 ft. An *external* bonding jumper, as shown here, is permitted, provided that the length *of the jumper* is not over 6 ft.

circuit—any length of flex that would make the total series length over 6 ft. would have to use an internal or external bonding jumper, regardless of any other factors.

2. Any length of standard flex that would require a bonding jumper longer than 6 ft. may not use an external jumper. In the Code sense, when the length of such a grounding conductor exceeds 6 ft., it is *not* a *bonding jumper but* is *an equipment grounding conductor and must be run only* inside *the flex, as required by Sec. 250-57(b).* Every length of flex that is over 6 ft. must contain an equipment grounding conductor run *only* inside the flex.

Exception No. 1 of Sec. 350-5 says that an equipment grounding conductor (or jumper) must *always* be installed for a length of metal flex that is used to supply equipment "where flexibility is required," such as equipment that is not fixed in place. Exception No. 1 actually modifies Exception No. 2, which describes the conditions under which a 6-ft. or shorter length of metal flex (Greenfield) may be used for grounding through the metal of its own assembly, without need for a bonding wire. Because experience has indicated many instances of loss of ground connection through the flex metal due to repeated movement of a flex whip connected to equipment that vibrates or flex supplying movable equipment, the Exception requires use of an equipment bonding jumper, either inside or outside the flex, in all cases where vibrating or movable equipment is supplied—for assured safety of grounding continuity. The rule applies to those lengths of 3 ft. or less that are permitted by Sec. 350-4, Exception No. 2, because "flexibility is necessary."

Liquidtight metal flex

Sec. 351-9, Exception No. 2 of the NEC and the UL's *Construction Materials Director* (the Green Book) note that any listed liquidtight flex in 1¼-in. and smaller trade size, in a length not over 6 ft., may be satisfactory as a grounding means through the metal core of the flex, without need of a bonding jumper (or equipment grounding conductor) either internal or external (Figure 3). Liquidtight flex in 1¼-in. and smaller trade size may be used without a bonding jumper inside or outside provided that the "total length" of the flex "in any ground return path" is not over 6 ft. Thus, two or more separate 6-ft. lengths installed in a raceway run would not be acceptable with the bonding jumper omitted from all of them. In such cases, one 6-ft. length or more than one length that does not total over 6 ft. may be used with a bonding jumper, but any additional lengths 6 ft. or less in the same

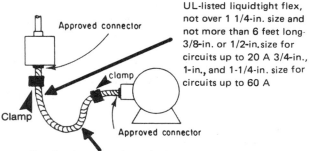

Approved connector

UL-listed liquidtight flex, not over 1 1/4-in. size and not more than 6 feet long- 3/8-in. or 1/2-in. size for circuits up to 20 A 3/4-in., 1-in., and 1-1/4-in. size for circuits up to 60 A

clamp

Clamp

Approved connector

Bonding jumper **not** required — metal in liquidtight is suitable for ground continuity

Figure 3   Watch out! Although *both* the UL rules on liquidtight flexible metal conduit and the rules of Sec. 351-9, Exception No. 2 clearly recognize applications of such liquidtight *without* need for an equipment grounding conductor—as suggested in this sketch—the application shown here would be a clear Code *violation* if clamps were omitted within 12 in. of the flex ends. Those clamps may be omitted *only* "at terminals where flexibility is necessary"—as indicated in Exception No. 2 of Sec. 351-8. *But* Sec. 351-9, Exception No. 1, makes it mandatory to use an equipment grounding conductor "where flexibility is required." The net effect is this: If you don't use the clamps you must use a ground wire. If you do use the clamps, the grounding conductor (the bonding jumper) is *not* required.

CLAMP SPACING on this liquidtight flex almost satisfies the basic rule of Sec. 351-8 that they be within 12 in. of each termination. It might be argued that such an arrangement does not constitute a condition where "flexibility" is necessary (that is, it does not utilize Exception No. 2 of Sec. 351-8); and, as such, it is *not* necessary to use an equipment grounding conductor with the flex from the switch to the A/C unit, because the size of the UL-listed flex and the 20A protection on the circuit satisfy Exception No. 2 of Sec. 351-9. *Avoid any problems:* always use an equipment grounding conductor (inside or outside the flex).

raceway run must have an internal or external bonding jumper sized from Table 250-95.

The required conditions for use of liquidtight flex without need of a separate equipment bonding jumper (or equipment grounding conductor) are as follows: (Beware! Later on, there's a twist to this.)

Where terminated in fittings investigated for grounding and where installed with not more than 6 ft. (total length) in any ground return path, liquidtight flexible metal conduit in the ⅜- and ½-in. trade sizes is suitable for grounding where used on circuits protected at 20A or less, and the ¾-, 1-, and 1¼-in. trade sizes are suitable for grounding where used on circuits protected at 60A or less.

The following are not considered to be suitable as a grounding means:

1. The 1½-in. and larger trade sizes.
2. The ⅜- and ½-in. trade sizes where used on circuits protected by fuses or CB rated higher than 20A, or where the total length in the ground return path is greater than 6 ft.
3. The ¾-, 1-, and 1¼-in. trade sizes where used on circuits protected at more than 60A, or where the total length in the ground return path is greater than 6 ft. Although UL gives the same grounding recognition to their "listed" liquidtight flex, this Code rule also applies to liquidtight that the UL does not list, such as high-temperature type.
   a. For liquidtight flex over 1¼ in., UL does not list any as suitable for equipment grounding, thereby requiring use of a separate equipment grounding conductor installed in *any* length of the flex, as required by Code. If a length of liquidtight flex larger than 1¼ in. is short enough to permit an external bonding jumper not more than 6 ft. long between external grounding-type connectors at the ends of the flex, an external bonding jumper may be used. *But watch out!* The rule says the *jumper,* not the flex, must not exceed 6 ft. in length *and* the jumper "shall be routed with the raceway"—that is, run along the flex surface and not separated from the flex.
   b. If any length of the flex is *over 6 ft.,* then the flex is not a suitable grounding conductor, regardless of the trade size of the flex, whether it is larger or smaller than 1¼ in. In such cases, an *equipment grounding conductor* (not a "bonding jumper," the phrase reserved for short lengths) must be used to provide grounding continuity and *it must be run inside the flex, not external to it.*

When a bonding jumper is required—such as for a length of the flex that is not over 6-ft. long but is over 1¼-in. size—Sec. 351-9 permits internal or external bonding of liquidtight flex as covered previously for standard flex. But for *any* size of flex run over 6 ft., *only* an internal equipment grounding conductor will satisfy.

Extremely important!

As described above for both standard metal flex and liquidtight metal flex, there are applications in which it is *not* necessary to run an equipment grounding conductor (an internal or external "bonding" conductor for flex lengths not over 6-ft. long). But great care must be taken to observe Exception No. 1 in Sec. 350-5 and Exception No. 1 in Sec. 351-9. Both of these rules—one covering metal flex and the other covering liquidtight metal flex—*warn* that an equipment grounding conductor *must always* be used with any piece of either type of flex if the flex connects to equipment "where flexibility is required."

In Sec. 350-5 on grounding of standard flexible metal conduit, Exception No. 1 says that "a grounding conductor shall be installed" with the flex for *every* application where the flex is "used to connect equipment where flexibility is required." And Exception No. 1 of Sec. 351-9 makes the same requirement for liquidtight flexible metal conduit. Those two rules were added in these sections because there were many reports, from field experience, that when either type of flex is used to supply equipment that is movable or subject to vibrations, the movement or vibration frequently causes separation of the conduit end from the connector to the equipment. When that happens, the equipment ground-fault current return path is broken if the only equipment grounding conductor is the metal of the flex. By insisting on a ground wire in (or outside of) any flex subject to "flexibility," the ground path will not be lost due to accidental opening of the connector.

The phrase "where flexibility is required" can be taken to mean any application where there is not a clamp on the flex within 12 in. of each end of the flex. This idea was derived from both Sec. 350-4, Exception No. 2 and Sec. 351-8, Exception No. 2. Each of those Exceptions was made to the basic support rule of Sec. 350-4 and 351-8, which requires clamping (support) of either flex at intervals of not over 4½ ft. *and within 12 in. of each end of a flex run.* Exception No. 2, in each case, permits omission of these clamps on flex lengths not over 3-ft. long "at terminals where flexibility is required." So, if the need for "flexibility" is the reason used to omit the clamps, then an equipment grounding conductor *must* be used with any such unclamped length of flex because it is being used "to connect equipment where *flexibility* is required."

Conversely, it might be argued that, if any length of flex is clamped within 12 in. of each end, the flex is *not* being used where "flexibility is required" and—therefore—an equipment grounding wire might be omitted if the length of the flex or liquidtight is not over 6 ft. long and if the conductors within the flex are supplied from a circuit protected at not more than 20A or 60A, under the conditions given in Exception No. 2 of either Sec. 350-5 or Sec. 351-9—exactly as shown in Figure 3.

It is an extremely common occurrence that vibrations or impact or some other physical action causes flex to separate from the flex connector body at equipment. If grounding continuity is provided solely by the metal of the flex length itself, opening of the type shown breaks the equipment grounding path and sets up the chance that an insulation failure (ground fault) in the equipment supplied will put a dangerous potential on the metal equipment enclosure—with consequent shock- and electrocution hazards. In both Sec. 350-5 and Sec. 351-9, the Exception No. 1 has the effect of requiring use of a separate equipment grounding conductor in *all* applications of flex that are not clamped within 12 in. of each end of the flex run. Thus, if separation occurs at a connector, the equipment ground path remains intact and a ground fault in the equipment will result in enough current flow to trip the CB or blow the fuse.

For all of the zigs and zags these Code rules take with respect to elimination of the need for an equipment grounding conductor with flex, it is our firm conviction that the complex concern is best resolved in favor of *total safety* (and minimum chance for accidents) by following the precept: *Use an equipment ground wire with* every *flex connection to equipment.* The extra labor and material expenses are low-cost insurance protection against potentially serious problems.

The *one* application where it seems very safe to accept Code permission for eliminating equipment grounding conductor is the widely used "fixture whips" for recessed incandescent fixtures where a 4–6-ft. length of ⅜-in. or ½-in. flex is used to satisfy the rules of Sec. 410-67(c). Figure 4 shows an application of that rule, as follows.

Recessed fixtures are, in all cases, marked with the required minimum-temperature rating of wiring supplying the fixture. Where a fixture wiring compartment operates so hot that the marked temperature exceeds the temperature rating of the branch-circuit conductors, high-temperature fixture wires must be run to the unit. The circuit outlet box supplying the high-wattage incandescent fixture must be mounted not less than 1 ft. away from the fixture. The flex whip may be ⅜-in. flex, for the number and type of fixture wires as specified in Table 350-3 (Sec. 350-3). If the branch-circuit supplying the fixture is protected at 20A or 15A, No. 18 fixture wires may be used for fixture loads up to 6A or No. 16 fixture wire may be used for loads up to 8A (Sec. 402-5), and the metal flex *may serve* as the sole equipment grounding conductor [Sec. 250-91(b), Exception No. 1]. The

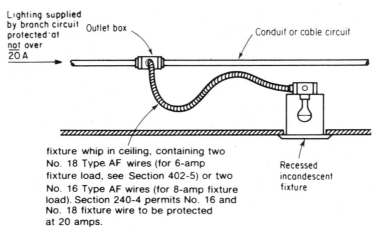

Lighting supplied
by branch circuit   Outlet box
protected·at   Conduit or cable circuit
not over
20 A

fixture whip in ceiling, containing two
No. 18 Type AF wires (for 6-amp   Recessed
fixture load, see Section 402-5) or two   incandescent
No. 16 Type AF wires (for 8-amp fixture   fixture
load). Section 240-4 permits No. 16 and
No. 18 fixture wire to be protected
at 20 amps.

Figure 4   This is the basic idea of "fixture whip," as covered in Sec. 410-67(c). The length of the flex (or liquidtight flex) must be between 4 and 6 ft., with the circuit outlet box located at least one foot from the fixture. The flex requires no clamping (Exception No. 3 to both Sec. 350-4 and Sec. 351-8). An equipment grounding conductor is not required for the flex if the circuit supplying the fixture is rated not over 20A.

FIXTURE WHIP is the wiring arrangement described in **NEC** Sec. 410-67(c), which requires suitably high temperature rating of the tap conductors that feed into the hot junction box on top of an incandescent fixture. The hookup shown here violates the **Code** rule because the junction box from which the circuit to the fixture is tapped is too close to the fixture and, therefore, subject to high heating radiated by the hot fixture. Each branch-circuit junction box is required by Sec. 410-67(c) to be positioned at least one foot away from the fixture. If the circuit feeding the fixtures through the junction boxes are protected by overcurrent devices rated not over 20A, there is no need for an equipment ground wire to be run with the flex.

flex may not be less than 4-ft. long but not more than 6-ft. long. The fixture wires could be Type PF for conditions requiring a 150°C rating or could be another type of adequate temperature rating for the fixture's marked temperature, selected from Table 402-3.

The 4- to 6-ft. fixture "whip" may be Type AC or Type MC cable, if the cable contains conductors of adequate temperature rating for the temperature of the fixture wiring compartment. This is an alternative to using a 4- to 6-ft. length of flexible raceway to enclose the tap conductors of the whip.

The purpose of this requirement of Sec. 419-67(c) is to allow the heat to dissipate so that heat from the fixture will not cause an excessive temperature in the outlet box and thus overheat the branch-circuit conductors which could be of the general-use type limited to 60°C or 75°C temperatures. This rule does not apply to "prewired" fixtures designed for connection to 60°C supply wires.

We've wrestled with these "zigs and zags" in court litigation, where the electrical defendant's reputation and future are on the line. It's no joke!

1. Both flexible metal conduit and liquidtight flexible metal conduit are primarily (almost exclusively) used for making short raceway connections (up to 6 feet) to supply motors, transformers and other equipment where some degree of flexibility is desired because the equipment supplied is not fastened in place, or has some movement or moveability in use, or where isolation from vibration is needed.

2. The NEC says any run of flex must be clamped within 12 inches of each end of the flex (Sec. 350-4 and 351-8). But if the flex is being used "at terminals where flexibility is necessary" (which is almost always why it is being used), then the flexibility may be achieved by omitting a clamp within 12 inches of the flex end, and up to a 3 foot length may be used without clamping.

3. Both UL and the NEC permit the metal construction of up to a 6-foot length of flex to be used as the equipment grounding conductor for the circuit run through the flex—with certain limitations on the maximum rating of the overcurrent device protecting the circuit (Exceptions No. 2 to Sec. 350-5 and Sec. 351-9).

4. But then Exceptions No. 1 in both Sec. 350-5 and Sec. 351-9). say that if the flex is being used "to connect equipment where flexibility is required"—which is virtually always the reason we're using it—an equipment grounding conductor (some call it "an equipment bonding jumper") must *ALWAYS* be used.

Cables and Raceways Through Framing Members

Sec. 300-4. **Type AC cable run *through* (i.e., perpendicular to) *metal* framing members do *not* require supplemental protection.**

Question: I have an installation where Type AC (BX) is run through factory-punched holes in metal framing members. This part of the installation was "red-tagged" because the inspector felt that additional protection in the form of a steel kick plate, as required in various parts of Sec. 300-4, was necessary to protect the armored cable. I disagree. I feel that such protection is *not* required. Am I right?

Answer: Yes. Secs. 300-4(a)(1) and (2), as well as Sec. 300-4(d) give the protection requirements for cables and raceways through *wood* framing members, either bored holes or notches, and for cables and raceways *parallel* to either *wood* or *metal* framing members, respectively. For these two applications, it is agreed that some form of additional protection—such as a steel sleeve, steel kick plate, or steel clip not less than ¹⁄₁₆ in. thick—is required for *all* cables and raceways, except those raceways specifically mentioned in the Exception to each of these rules—i.e., rigid steel conduit (Art. 346), IMC (Art. 345), rigid nonmetallic (Art. 347), and EMT (Art. 348). All other cables and raceways must be provided with the additional protection required by these rules.

However, Sec. 300-4(b), which is titled "Cables and Electrical Nonmetallic Tubing Through Metal Framing Members," is the section that deals with the Code-required protection where a "cable" runs *perpendicular* (not parallel) through *metal* (not wood) framing members. This rule would be the one that covers the application described above and it only requires that the additional protection spelled out in part (2) of Sec. 300-4(b) be provided where "*nonmetallic sheathed cable* or *electrical nonmetallic tubing*" passes through a metal framing member. Inasmuch as Type NM cable and ENT are the only two methods referred to, and there is no mention of any other cable or raceway, it is reasonable to conclude that the additional protection (steel plate, etc.) would *not* be required to protect Type AC (BX) cable where it is run through (and perpendicular to) *metal* framing members.

Article 305 of the NEC

Secs. 305-4 and 305-6. For temporary wiring on construction sites, the specific requirements of Article 305 must be observed.

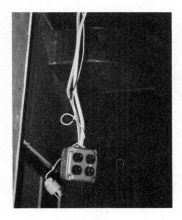

Article 305 of the NEC covers temporary wiring for light and power for the various trades working in a building under construction. Because of the many safety factors involved, great care must be exercised. OSHA rule 1910.304(b) covers "Ground-fault protection for personnel on construction sites"—requiring that receptacles have *either* GFCI protection (by being GFCI receptacles or protected by GFCI circuit-breaker at the panel) or be provided with OSHA's "assured equipment grounding conductor program." Use of receptacles in a box connected to 2 Romex cables by a single box clamp (as shown in this photo) with the box simply suspended at the end of the cables, which are hung from an overhead taped connection to another cable, would pose potential liability in case of injury to personnel. And the plug-cap on the end of the black and white wires from overhead is used to energize temporary light fixtures by plugging into a receptacle. NEC Sec. 305-4(c) would not accept the "open wiring" as shown and (d) says, "Receptacles on construction sites...shall not be installed on branch circuits which supply temporary light." Carelessness in temporary wiring can be a ticking time bomb.

Note 8 Derating Factors

Sec. 310-15, and Note 8 to Tables 310-16 through 310-19. Derating factor where more than three current-carrying conductors are installed in a raceway or cable must be taken from either Column A or Column B of the table to Note 8.

Editor's Note: The following question is answered by Mr. John M. Caloggero, Senior Electrical Field Service Specialist with the National Fire Protection Association (NFPA) and co-editor of the NFPA's 1990 NEC Handbook. Because the question involves commentary from the NFPA's Handbook, we felt it only proper that the answers come from the editors of that publication. Although Mr. Caloggero is an employee of the NFPA, as pointed out at the end of his answer, this is not a formal interpretation, but rather his opinion on the topics discussed. Many thanks to John for his support of our effort to answer these questions.

Question: Please review the enclosed notes on ampacity adjustment factors as covered on pages 310 and 311 of the NFPA's 1990 NEC Handbook, and clarify the confusing points—especially those I have noted:

1. In the second paragraph, last sentence: "If the watt loss is equal to or less than 3062.5R, Column A correction factors can be used." If the watt loss is equal to 3062.5R, what other choice is there for one to use?

2. In the second paragraph, last sentence: "What is meant by 'Anything greater than Column B applies.'?"

3. On page 311, in the third paragraph, first sentence: "...with all conductors carrying full load (no diversity) can be calculated as follows: $A_1 = 25 \times 0.7 = 17.5$." How is this possible when the single asterisk note at the bottom of the table to Note 8 states: "These factors include a load diversity of 50 percent."

4. In the same paragraph, second sentence: "Table 310-16 ampacity multiplied by derating factor for Column A:

$$N = 20$$

$$E = 20$$

$$A_2 = \sqrt{0.5 \times 20/20} \times 17.5 = 12.37 \text{ amperes."}$$

Where is the derating factor for Column A located in the above formula? (The 0.5 appears to be indicating a 50 percent load diversity.)

5. What is the relationship between the formula in 4 above and the calculation in paragraph 3 on page 310?

6. When dealing with the single-asterisk (50 percent diversity) items in the table to Note 8, do we calculate per Column A and B, and then select the lower current value?

Mr. Caloggero's Answer: I will answer your questions in the order asked.

The concept used in applying Note 8 is based on maintaining a balance between the heat generated by current traveling through the conductor and the ability of the medium surrounding the conductor to dissipate the heat to maintain the temperature below the temperature rating of the conductor insulation. The percentage values in Column A were based on tests where there were no more than nine current-carrying conductors in raceway. Once you exceed nine conductors, you were required to apply the 50 percent diversity rule when applying the derating factors. Therefore, the maximum heat loss allowed is 3062.5R watts for any combination of ampere loads on the conductors. If the watt loss is equal to 3062.5R watts, for the example given, you have reached the maximum load current on the conductors. The options are to reduce the number of conductors in the raceway, reduce the load on the circuit, or use larger conductors.

In the last sentence of the second paragraph on page 310 there is a typo. The word *than* should be replaced with the word *then*.

In your question No. 4, A_1 used in the formula is the result of multiplying the Table 310-16 ampacity by the percentage factor in Column A.

Your fifth question asks for the relationship between this formula and the calculation made in the third paragraph on page 310. The third paragraph provides an example in which all twenty conductors are carrying 8.75 amperes. Their total heat loss is 1531.25R watts, which is less than the maximum allowed. If we apply the formula to the problem, the resulting value is 12.37 amperes as shown on page 311, which is approximately 50 percent of 25 amperes.

Where ten or more conductors are installed in a raceway or cable, Column A is used, keeping in mind the 50 percent load diversity as indicated by the single-asterisk note. There is no need to make a calculation in accordance with Column B unless you have not satisfied the 50 percent load diversity. Where the 50 percent load diversity is not complied with, Column B must be used.

I hope that the above personal comments serve to answer your questions. Please note that because the above personal comments have not been processed in accordance with the NFPA Regulations Governing Committee Projects, Section 16, they are not to be considered to be a "formal interpretation."—John M. Caloggero, Sr. Electrical Field Service Specialist, NPFA.

Editor's Comments: This whole business seems rather complicated because the rule of Note 8 simply states that the ampacity value shown in Tables 310-16 through 310-19 must be derated where there are more than three (current-carrying) conductors in a raceway or cable. And, as given in the wording of Note 8, the derating value "shall be selected from the following table." Therefore the only question is, "Which column do I use, Column A or Column B?"

The single-asterisk note to the Table of Note 8 indicates that where more than 10 conductors are in a raceway or cable, the derating factors shown are based on a 50 percent diversity. Although that term is not defined anywhere in the NEC, there is some indication in the available documentation that "a 50 percent diversity" means that half the conductors within a given raceway or cable are carrying a value of current that was derated in accordance with Column A of the Table—based on the total number of conductors—and the other half of the conductors run in the raceway or cable are carrying no current.

The question from a practical standpoint is: Where would Column A ever be useful? If the ampacity of each individual conductor is *not* based on the idea that any "non-current-carrying" conductors will eventually be loaded, then the "non-current-carrying" conductors may *never* carry current without recalculating the ampacity of the "current-carrying" conductors, and, if necessary, changing the loading and/or the rating of the overcurrent protective device. Additionally, if the "non-current-carrying" conductors are never to carry current, then why are they there? How is it possible to prevent these "non-current-carrying" conductors from being used at a future date?

In addition to the practical concerns, as indicated by the single-asterisk note at the bottom of the Table to Note 8, the derating factor of Column A may be used when there is a diversity of 50 percent, that is, if half the conductors are carrying no current, then the derating factor for the other half of the conductors may be taken directly from Column A. What if the split is not exactly 50/50? Is it permissible to use the derating factors given in Column A?

There has been much commentary published with regard to that question. Some contend that the formula given in the Fine Print Note (FPN) can be used in conjunction with the derating values given in Column A and a derating factor that will result in an equivalent heating effect can be developed. That, however, does not comply with the basic requirement which states: "Where the number of conductors in a raceway or cable exceeds three, the ampacities *shall be reduced as shown in the following table:*"

The word *shall* indicates a mandatory Code requirement. Additionally, the formula in the Fine Print Note is strictly informational and presents no rule. If the formula given in the FPN were an exception to the basic rule—as was originally proposed—then use of the formula to determine a derating factor based on "equivalent heating" would be acceptable. As it now stands, the basic rule (i.e., selecting a derating factor from the *Table*—Column A or B) must be satisfied.

Because application of the derating factors from Column A will result in greater heating of the conductors if more than half of them are carrying that derated value of current, the factors from Column A may *not* be used unless 50 percent or less of the conductors are carrying current. To comply with the basic rule, if I can't use Column A, then I must use Column B.

Although some call it "the coward's way out," the easiest and surest way to comply with the requirements for derating of conductor ampacity where more than three current-carrying conductors are in a raceway or cable, is to always apply the factors shown in Column B. That way, loading and protection of *all* conductors can be determined and, to some extent, fixed at the design stage. This will provide for maximum use of the conductors without having to constantly recalculate the conductors' ampacities as the "extra" conductors are loaded. Additionally, it should assure that the "current-carrying" conductors do not become overloaded where the ampacities are *not* recalculated as load is added.

Derating of Underground Duct Banks: A Tricky Task

Secs. 310-15(a), 310-15(b), Tables 310-69 through 310-84, and Appendix B. De-termination of required derating for underground circuits in "duct banks" is not clearly spelled out where the "optional" method for establishing conductor ampacity (up to 2000V) is used or where more than six ducts are used with circuits rated 2001 to 35,000V.

Design of underground circuits in duct banks must provide effective ampacity derating for conductors that are run in ducts or conduits, particularly in the middle of a duct bank, where the I^2R heating is less effectively dissipated than in the outer or perimeter ducts. The reality is that the interior conductors in ducts are surrounded by other conduits that are heated by their internal conductor losses, and must be used at lower ampacity loading. But the difficult design determination is to determine what ampacity derating factors must be applied, based on the total number of ducts and the insulation temperature rating of the installed conductors. And this concern applies to *all* duct banks, regardless of the voltage rating of the circuits—to circuits up to 600 volts, as well as to 5-kv, 15-kv, and 35-kv circuits.

The 1990 National Electrical Code does offer some guidance in the task of assigning safe and reliable ampacities to conductors in duct banks. To begin, Sec. 310-15 does clarify that the phrase "electrical duct(s)" as used in the NEC includes "any of the electrical conduits recognized in Chapter 3 as suitable for use underground." That includes rigid metal conduit, intermediate metal conduit (IMC), EMT and conduits of fiber, asbestos-cement, soapstone, rigid PVC, fiberglass epoxy, and high-density polyethylene. And the phrase "electrical duct(s)" will include any of the conduits whether they are "embedded in earth," without any concrete encasement, or encased in concrete in the earth.

The NEC rules on ampacity of underground duct circuits can be divided into two categories—those applying to circuits rated up to 2000 volts and those rules that apply to circuits rated 2001 up to 35,000 volts, as follows:

Up to 2000 volts: For circuits derived from the common electrical systems rated up to 2000 volts—120/240V, single-phase, 3-wire; 208/120V, 3-phase, 4-wire; 240V, 3-phase, 3-wire; 120/240V, 3-phase, 4-wire, high-leg delta; 480V, 3-phase, 3-wire; 480/277V, 3-phase, 4-wire; and 550V, 3-phase, 3-wire—the mandatory limitations on con-

ductor ampacities in underground ducts are those contained in Sec. 310-15(a) and Tables 310-16 and 310-18, with the Notes to those Tables. In effect, for any circuits rated up to 2000 volts, with conductor temperature ratings up to 90°C—such as THW, THWN, RHW, THHW, and XHHW—the ampacity values determined from Table 310-16 and its notes would apply to conductors in *any* underground ducts, regardless of the number, spacing, and configurations of the ducts. Table 310-16 (as well as Table 310-18 for conductors with insulation temperature ratings from 150°C to 250°C) applies to *any* conductors in raceway—underground as well as above ground—without any concern for heat accumulations among closely placed conduits and without any difference in conductor ampacities based on number or placements of conduit runs in duct banks.

For circuits up to 2000 volts, that simple approach to determining conductor ampacities has been used for decades and has generally proved safe and effective. The acceptability of such practice has been attributed to conservative ampacity values in Tables 310-16 and 310-18, along with only very limited application of large banks of multiple ducts for circuits at voltages below 2000 volts. However, the 1990 NEC does offer *nonmandatory* guidance on more precise conductor current loadings for underground duct banks up to 2000 volts.

Sec. 310-15(b) of the NEC is intended to permit a more detailed and heavier current loading of conductors in duct banks, but only in accordance with application data given in "Appendix B" on page 781 of the NE Code book. For example, under the basic conditions of Table 310-16, 1 No. 1/0 THW copper conductor in an underground duct has an ampacity at 150A, but under the conditions of Table B-310-7, a No. 1/0 THW copper conductor may have any one of nine ampacities varying from 111A up to 197A—depending upon the number of ducts in the bank (1, 3, or 6) and for varying values of RHO [thermal resistivity of soil, as described in Sec. B 310-15(b)(2) in Appendix B].

As stated in the first sentence of Appendix B, "This Appendix is not part of the requirements of this Code, but is included for information purposes only." All the data and procedures presented are, therefore, optional guides to establishing conductor ampacities. Secs. B 310-15(b)(1) to B 310-15(b)(7)—along with Tables B 310-1 to B 310-10 and Figures B-310-1 to B-310-5—present an ampacity determination procedure that applies to underground-duct circuits operating up to 2000V. Table B 310-7 covers single insulated conductors, three conductors per duct, in underground electrical ducts and can be considered the basic alternative to Table 310-16, with all the conditions and modifications described in Appendix B.

It should be noted that the ampacity value determined from Appendix B for any given conductor in a duct bank applies for all of the conductors in the duct bank—without any differentiation between conductors in the middle of the bank and conductors at the perimeter of the bank cross-section. As indicated later, there should be loading differences based on duct positions within a bank (see Figure 1).

2001 to 35,000 volts: Sec. 310-15(a) makes it *mandatory* to calculate ampacity of medium-voltage circuits in accordance with Tables 310-69 through 310-84 and their accompanying notes. Table 310-77 is generally the medium-voltage equivalent to Tables 310-16 and B 310-7 for circuits up to 2000 volts—covering three single insulated copper conductors in underground electrical ducts (three conductors per electrical duct) for 90°C conductor insulation.

For medium-voltage circuits, ampacity determinations from Tables 310-69 through 310-84 must satisfy the data given in the Notes on page 183 of the NE Code book, along with the data in Figure 310-1 on duct bank arrangements. Tables 310-77 to 310-80, covering underground circuits in ducts, refer to "Detail 1," "Detail 2," and "Detail 3" of Figure 310-1—for duct banks of one, three, or six ducts, as shown in Figure 310-1. Those Tables show successively lower allowable conductor ampacities as the number of ducts increases. For instance, a 250 kcmil copper conductor goes from 321A to 260A to 210A—to accommodate the higher heating due to reduced dissipation as the concen-

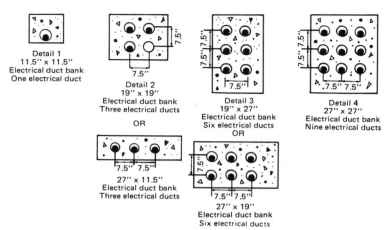

Figure 1   These are underground duct configurations shown in NEC Fig. B 310-2, covering circuits up to 2000 volts and Fig. B 310-1, for circuits rated 2001 up to 35,000 volts. These details—1, 2, and 3—are cross-referenced to the ampacity Tables 310-77 through 310-84 and Tables B 310-6 and B 310-7. Detail 4 is not referenced, but Fig. B 310-3 in Appendix B of the NEC book presents design information on that 9-duct configuration.

tration of ducts increases. *But,* the ampacity Tables make no reference to "Detail 4" in Figure 310-1, showing nine electrical ducts. For any applications over six electrical ducts, additional reduction in allowable ampacities must be made by extrapolating the rate of ampacity reductions shown for one to six ducts in any of the four underground duct Tables.

As noted above, the NEC rules and recommendations on ducts apply to underground duct banks that are *either* directly grouped in earth or enclosed by a concrete envelope. For that reason, the duct "Details" shown in Figure 310-1 must be taken as applying with or without the concrete shown.

Important!

Note 3 of the "Notes to Tables 310-69 through 310-84" *requires* that conductor ampacities in duct banks must account for differing heating conditions between "inner electrical ducts" and "outer electrical ducts." Although this consideration is not covered in Appendix B for circuits up to 2000 volts, design should properly account for the difference. Note 3 reads as follows:

3. Electrical Duct Bank Configuration   Ampacities shown in Tables 310-77, 310-78, 310-79, and 310-80 shall apply only when the cables are located in the outer electrical ducts of the electrical duct bank. Ampacities for cables located in the inner electrical ducts of the electrical duct bank shall be determined by special calculations.

That last phrase referring to "special calculations" leaves this important concern totally to the designer. Figure 2 shows information made available in an old study of this matter.

Effect of Cable Position in Duct Bank

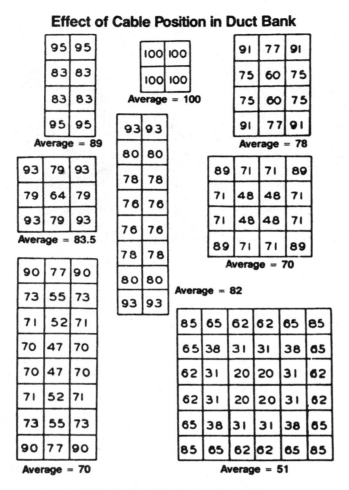

Relative Watt losses for individual ducts and bank of ducts per foot of cable for same temperature rise (per cent of 2 by 2 bank).

$$\text{Position Factor} = \sqrt{\frac{\text{Relative Loss for any Duct}}{\text{Average Loss}}}$$

Figure 2   When several equally loaded cables are placed in the same duct bank, the maximum permissible loading is the root mean square peak load of all cables. If all cables actually carried this load, the operating temperature of those in interior ducts would exceed the safe operating temperature of the insulation. Cables in the outer ducts would not reach this value. Overloading and underloading respectively would occur. To equalize the operating temperatures of cables in the various locations at a value corresponding to the safe maximum temperature, the load must be multiplied by a position factor. This will decrease the load for the inner ducts and increase it for the outer ducts. The root mean square of the loads remains unchanged. Position factors for several arrangements of ducts may be obtained from data published by P. Torchio (A.I.E.E. Transactions, 1921). For unequally loaded cables, the same load position adjustment should be made.

Locking a Switch in the "Closed" Position Is *Not* Required by the NEC

Secs. 380-3, 380-4, and 380-8. **Lawsuit focuses on NEC rules regarding locking of switches and enclosures.**

A sewer-construction contractor retained the services of an electrical contractor to install a 480VAC temporary outdoor service to supply the electrically-powered underground, earth-boring machine used to install sewer lines. On a weekend, when work was not in progress on the job site, some unknown person opened the outdoor service disconnect switch. One of the electrically-powered equipments fed by the service was a sump pump, which was installed to prevent water accumulation (flooding) in the underground cavity where the million-dollar boring "mole"—a 60hp motor-driven excavating machine—was situated. When heavy rain and normal surface water resulted in flooding of the underground cavity, the pump could not operate and very costly damage was done to the "mole." The attorneys for the sewer contractor's insurance company filed suit against the electrical contractor, alleging that his design of the service was faulty in not *locking* the service disconnect in the *closed* position, to prevent unauthorized opening of the service disconnect, to thereby assure that the sump pump would always be operative and prevent flood damage to the expensive "mole."

OFFICE OF THE SHERIFF

  County

CIVIL ACTION

𝔖𝔲𝔪𝔪𝔬𝔫𝔰

The complaint

The following are verbatim *excerpts* from the court papers and Summons presented to the electrical contractor:

First Count

1. At all times hereinafter mentioned, the Plaintiff was installing sewer piping along XXXXXX Avenue in XXXXXX, XX.

2. On or about December XX, XXXX, Defendants, XXXXXXX, negligently furnished and installed a temporary 480V electrical service on XXXXXX job site in XXXXXXX, so as to create an attractive nuisance. The enclosure housing for the electrical service was secured by tape rather than a lock.

4. On or about December 27, XXXX, an unknown person cut the tape and turned off the electrical service to the Plaintiff's job site and the Plaintiff's equipment and tools were damaged by water which accumulated on the job site when the sump pump used to pump water out of the site stopped as a result of the electrical service being turned off.

Second Count

2. The Defendants, its agents, servants and/or employees, performed its services in a negligent manner in that they:

A. Failed to properly secure the enclosure housing for the electrical service in a proper manner;

B. Failed to properly supervise its agents, servants and/or employees;

C. Failed to perform the work in a proper workmanlike manner;

D. Committed other acts of negligence.

Third Count

2. The Defendants expressly or impliedly warranted said electrical housing as reasonably fit, suitable or safe for its intended purpose. The Defendants breached their implied warranty of merchantability pursuant to XXXX 12A:2-314 and their implied warranty of fitness for a particular purpose, pursuant to XXXX 12A:2-315.

Fourth Count

2. Pursuant to XXXXX 2A:58C-2, the Defendants are liable to the Plaintiff in that the produced, designed, manufactured and/or distributed by the Defendants, which caused the Plaintiff's damage, was not reasonably fit, suitable or safe for its intended purpose in that it:

A. Deviated from the design specifications, formulae or performance standards of the manufacturer to the same manufacturing specifications or formulae;

B. Failed to include adequate guards, warnings, safeguard and/or instructions;

C. Was otherwise designed in a defective manner.

For All Four Counts

As a direct and proximate result of the Defendant's actions on or about December 27, XXXX, the Plaintiff's electrical service was turned off and the sump pump stopped operating, causing extensive damage to the Plaintiff's equipment and tools. The Plaintiff has incurred expenses in an effort to repair the damages so sustained.

WHEREFORE, Plaintiff demands judgment against the Defendants for damages, together with interest, attorney's fees and cost of the suit.

Jury Demand   Plaintiff hereby demands trial by jury as to all issues.

Analysis

Figure 1 shows a simplified layout of the outdoor arrangement for service to the switch in a weatherproof enclosure. And the accompanying photo shows a very commonly found installation of outdoor switches that are accessible to the general public but do not have, and are not required to have, any locking provisions on the operating handle.

The first question that must be answered is, "Does the NE Code re-

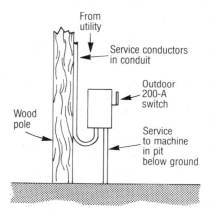

Figure 1

Outdoor switches shown here are accessible to passersby and the general public, without any provision on their handle to prevent unauthorized operation or vandalism.

quire that such a service disconnect be *locked* in the *closed* position under any conditions?"

Answer: The NE Code has no rules requiring a service disconnect to be locked *closed,* or even locked *open.* Article 230 on "Services" has no such rule.

Article 380 on "Switches" requires a switch to be externally operable and mounted in an enclosure listed for the intended use (Sec. 380-3). Sec. 380-4 requires use of a weatherproof enclosure for a switch installed in a "wet location"—such as outdoors exposed to the rain and weather. *But,* there are no rules in Article 380 that *require* a switch to be locked open or locked closed. And there are no rules even requiring that such a switch be *capable* of being locked closed or open—such as would be provided by a locking hasp on the external handle assembly that controls the "open" and "closed" position of the switch contact.

Sec. 380-8 does require that a switch be located so that it may be operated from a readily accessible location. The Code definition for "readily accessible," as given in Article 100, simply requires that the switch must be capable of being reached quickly for "operation, renewal, or inspections"—but such accessibility has to be provided only for "those to whom ready access is requisite." The service switch in the installation under discussion here complies with all of the NEC rules on accessibility. And NEC 240-24 requires that fuses in such a switch must also be "readily accessible."

Sec. 110-17(a) of the NEC does require that live parts of electrical equipment (such as fuses and terminals in the switch enclosure) be guarded against accidental contact by being installed in an *enclosure* that is "accessible only to qualified persons." For the outdoor service switch, that requirement can be satisfied by providing a lock on the enclosure (but not on the handle), with authorized and qualified personnel having the key to the lock available to them. But that rule has nothing to do with the operating handle of the switch—which is *not* required to be *locked,* either in the open position or in the closed position.

"Is it ever a Code requirement to lock a switch handle in the closed (ON) position?" The answer is no. Then, "Is it a Code violation to lock a switch handle in the closed position?" The answer is again no. The simple fact of the matter here is that locking of a switch handle in *either* the closed *or* open position is a function of operation of the equipment with which the electrical parts are associated, and the National Electrical Code does *not* (and does not intend to) regulate "operating procedures" or even "safety-related work practices."

Precisely because the NEC does not cover that aspect of electrical equipment operation, The Occupational Safety and Health Administration (OSHA) of the U.S. Department of Labor has developed "work practices" to come to grips with nonelectrical injury or damage that may result from careless or ill-advised operations of electrical equipment (the so-called Lockout/Tagout Rules). Although the OSHA "Lockout/Tagout" rules were *not* completed or in effect at the time of this incident, it is worth noting that these rules place the responsibility for establishing and implementing the "Lockout/Tagout" procedures with the employer who owns manufacturing, processing, or production facilities that are electrically supplied and energized. The "work practices" are *not* the responsibility of the designer and/or installer of the electrical circuits and equipment that supply the employers' facilities. And, even so, OSHA's Lockout/Tagout rules never require a switch to be locked in the *closed* position.

The letter and intent of NEC Sec. 90-1 is that the NEC is concerned *only* with "hazards arising from the use of electricity." It is a long-established and well-understood position that the NEC seeks to protect *only* against hazards of shock, electrocution, fire, and explosion—as they may be caused directly by the electrical circuits and equipment. *But,* protection against machine and tool damage described in the court papers quoted previously is not within the purview of the NEC, which states in Sec. 90-1(b) that "compliance" with the Code will not necessarily make an installation "efficient, convenient, or adequate for good service."

The position of the NEC with respect to limiting its coverage solely to direct electrical hazards—and not ancillary or correlated hazards from equipment operation (or, as in this case, failure of the sump pump to operate)—is similar to the position taken by UL in its "listing" of electrical products. UL, for instance, is concerned that a TV set is essentially free of electrical hazards but *not* that it has a clear, sharp picture or even that it will perform its intended function properly or have an acceptable service life. UL says in its "General Information Directory" (the White Book), "Unless a specific safety function is concerned, the efficacy of a product has not been investigated."

Use of a padlock to lock a switch handle in the ON position (closed) does protect against unauthorized operation of the switch. And although this is not required by the NEC or any other electrical industry standards, it is also not a violation of any code or standard. Such provision is completely discretionary with the premises owner and is solely the owner's responsibility.

Padlocked enclosures for these switches prevent unauthorized access to energized parts and/or enclosed fuses—*but,* operation of the switch handle is not restrained by any locking provision.

Maximum Mounting Height for CBs "Used as Switches"

Secs. 240-81, 240-83(d), and 380-8(a). Circuit breakers used as switches must not be mounted such that the middle of the operating handle is more than 6-1/2 ft. above the ground.

Question: I have a residential installation where the 100A service panel is mounted inside the garage, on a wall, at a height of 6 ft. 8 in. above the floor, as measured to the top of the enclosure. My inspector refused to accept the installation because he said it was in violation of Sec. 380-8(a), which establishes 6 ft. 6 in. as the maximum mounting height for switches and CBs "used as switches."

I disagreed with my inspector because the CBs in the service panel are for circuit protection. They are not to be "used as switches" to control any loads, and therefore do not have to comply with the requirements of Sec. 380-8(a). Who's right?

Answer: As covered in Sec. 90-4 of the NEC, the final responsibility for interpreting the rules and requirements given in the NEC is the "authority having jurisdiction." And that "authority" is (typically) the local electrical inspector. Basically stated, whatever the inspector says, goes.

However, as you noted, the rule of Sec. 380-8(a) only applies to CBs "used as switches." And determination of whether or not compliance with the 6 ft. 6 in. maximum mounting height is mandatory for any application of CBs hinges on the interpretation of that phrase.

As we have all seen, there are a number of commercial and industrial installations where 120 to 277VAC lighting on a given floor or in a given area is switched ON and OFF at the branch-circuit panelboard that supplies the local lighting. This is permissible and recognized by the NEC provided the CBs are listed as suitable for switching duty and marked "SWD" [Sec. 240-83(d)]. In such an application, there is no question that Sec. 380-8(a) would apply and the CBs must be mounted no more than 6 ft. 6 in. above the floor—as measured to the center of the operating handle in its "highest position."

In the installation you describe, although it is not intended that *loads* will be switched ON and OFF at the service panel, what about when any of the *circuits* are switched ON and OFF for maintenance, repair, alterations, etc., using the branch-circuit CB? In that case, the CB is *not* performing a protective function as it does when it operates on short-circuit or overload. Instead it is being used to *switch* the cir-

cuit ON and OFF. Because virtually *every* CB is also "used as a switch" to disconnect the circuit it protects, inspectors will generally require compliance with the maximum 6 ft. 6 in. mounting height given in Sec. 380-8(a). Except for CBs used with busway installations, as a general rule, always ensure that CBs are not mounted more than 6-1/2 ft. above the floor or work platform.

You indicated in your question that the 6 ft. 8 in. height was measured to the top of the enclosure. The rule of Sec. 380-8(a), however, states that the distance to be considered is from the floor or working platform up to the "center of the grip of the operating handle...when it is in its highest position." (For vertically operated CB handles, that would be the ON position—see Sec. 240-81. For horizontally operated CB handles, either position will give the same height measurement.) In this case, if there is at least 2 in. of clearance between the top of the enclosure and the center of the top CB's or main CB's operating handle(s)—which there should be—then the installation *is* in compliance with the 6 ft. 6 in. maximum called for by Sec. 380-8(a). If not, maybe the offending breaker(s) could be relocated elsewhere within the panel, such as in vacant lower slots. If that is done, remember to close the unused openings.

THW in Ballast Compartments

Table 310-13 and Sec. 410-31.   Where used in ballast compartments of fixtures, THW has a 90°C rating.

Question: In Sec. 410-31 of the NEC, the Code requires that all conductors that come within 3 in. of a ballast in a fixture wiring-compartment have an insulation temperature rating of at least 90°C. The rule then goes on to give examples of the types of insulations permitted.

In the list of acceptable conductor insulations, THW is given. I thought THW is a 75°C wire and not 90°C. Is this a misprint?

Answer: No, this is not a misprint. Although in all other applications THW-insulated wire is considered to be a 75°C conductor, when used in a fixture wiring-compartment within 3 in. of the ballast, THW insulation has a 90°C rating. This is because THW has been found to be sufficiently capable of withstanding the elevated temperatures that exist within such close proximity to the fixture ballast.

This is covered in Table 310-13. It is worth noting that Table 310-13 also limits this permission to conductor sizes No. 14 through No. 8. And because Table 310-16 only gives the ampacity of THW-insulated conductors in the 75°C Column, the ampacity of THW conductors used in accordance with Sec. 410-31 must be based on the value of current shown in the 75°C Column, even if derating is to be applied.

Watch Out for "Wet-Location" Receptacles!

Sec. 410-57. Receptacles installed at "wet locations" must be provided with a cover that will protect the receptacle where supplying other than portable (temporary use) appliances or equipment.

Much public attention has been drawn to electrocutions of people who contacted energized outdoor metal enclosures of electrical equipment. Media reports of such deaths have been frequent and cover all parts of the nation. We reported on such a death involving a fault-energized metal lighting standard in New York City, on one involving an energized in-ground metal junction box in the mid-South and one on the West Coast. Because of liability suits arising from these unfortunate accidents, all outdoor electrical applications have become objects of great concern from the safety standpoint—especially because use of outdoor electrical equipment is continually increasing and because it all has such high visibility. Although effective grounding is the key to eliminating such problems, an even more basic need for all electrical designers and installers is to assure complete and strict compliance with *all* NEC and UL rules that are applicable. We must allow no compromises on selection and applications of outdoor electrical equipment, and that puts the spotlight directly on the most common outdoor electrical violation—receptacle outlets that do not satisfy NEC Sec. 410-57.

In our continual travels all over the nation, we are constantly amazed at the very widespread use of outdoor receptacles that are in violation of NEC Sec. 410-57 covering "Receptacles in Damp or Wet Locations." And in conducting Code courses, we constantly encounter confusion on the part of designers and installers about the relationship between NEC rules and UL rules. Let's take it step-by-step.

In Article 100, the NEC defines a "Wet Location" as one "subject to saturation with water or other liquids, such as...locations exposed to weather and unprotected." A "Damp Location" is an outdoor location that is "Partially protected...under canopies, marquees, roofed open porches, and like locations"—that is, not directly exposed to rain or other water or liquid. With respect to those same definitions, the UL White Book says "Boxes *and* covers intended for use in wet locations as defined by the NEC are marked "Wet Location." Damp location boxes *and* covers are marked "Damp Location."

Based on those two references—one in the NEC and one in the UL rules—it would seem logical for NEC Sec. 410-57 to require that "Receptacles in a damp location shall be in a box and cover that is *marked* and *listed* for damp locations." And "A receptacle in a wet location shall be in a box and cover that is *marked* and *listed* for wet location." But the wording of Sec. 410-57 is not at all that direct.

According to Sec. 410-57, a receptacle used in a damp location—such as an open or screened-in porch with a roof or overhang above it—may not be equipped with a conventional receptacle cover plate. It must be provided with a cover that will make the receptacle(s) "weatherproof" when the cover or covers are in place. UL rules make no reference to "weatherproof" and the NEC definition of that word is simply not clear. The type of cover plate that has a thread-on metal cap held captive by a short metal chain would be acceptable for damp locations but not wet locations. Any plate-and-cover assembly may be used in a damp location provided it covers the receptacle to make it "weatherproof" when not in use. The type of receptacle cover that has horizontally opening hinged flaps (doors) to cover the receptacles may be used in a damp but not wet location if the flaps are not self-closing, i.e., if the flaps can stay open. Of course, any cover plate that is listed for "wet location" use may also be used in damp locations and *must* be used if the receptacle damp location is subject to "beating rain" coming in at an angle or subject to "water run-off."

For wet locations, NEC Sec. 410-57(b) says receptacles must be used with either of two types of cover assemblies:

1. For a receptacle outdoors or in any other wet location where a plug-connected load is normally left connected to the receptacle—such as for outdoor landscape lighting or for constant supply to an appliance or other load—*only a UL-listed* "wet-location with plug-cap inserted" box and cover plate should be used. Such an assembly maintains weatherproof protection of the receptacle at all times—either with the plug out or the plug in. Such "wet-location" listed assemblies with a vertically lifting cover shields the receptacle against driving rain (coming at an angle) or hose spray. Cover assemblies use a vertically-lifting "canopy" that protects either a single or duplex receptacle.

2. For outdoor receptacles used solely for occasional connection of portable tools or appliances (lawn mowers, hedge-trimmers, etc.), it is permissible to use a cover that only provides protection against weather when the cover is closed (but not when a plug is inserted). But such a cover, whether installed for vertical or horizontal movement of the cover, must have spring-loaded self-closing covers or gravity-close for vertical-lift covers.

To minimize liability exposure always use outdoor receptacles with boxes and covers that are UL-listed and marked for "Wet Location"—while in use, with the plug connected.

NEC Motor Control-Circuit Rules:
Ridiculous Complexity Thwarts Safety!

Sec. 430-72(b). **This is an absolute fact: The rules of NEC Sec. 430-72(b) covering overcurrent protection for coil-circuit conductors of magnetic motor starters are so complex and confused that virtually all electrical design and installation people cannot fully understand them—or they simply ignore them because they are beyond enforcement. That's what they tell us.**

As we continually travel all over the nation—constantly conducting NE Code seminars, conferences, and workshops for people at every level of competence on Code rules—we are convinced that we spend more time than anyone else trying to analyze and understand Sec. 430-72(b), especially with engineering staffs of large industrial facilities. We find that the whole issue of motor-starter coil circuits is in great disarray. And that widespread reality has proved to be a very serious liability to electrical people in lawsuits resulting from injuries and fatalities involving motor drives—like conveyors, cranes, hoists, and machine tools.

The basic problem comes down to this: When must overcurrent protection (fuses) be installed at a magnetic motor starter to protect the coil-circuit wires that are run from the starter to a remote manual or automatic pilot-control device—such as a START-STOP pushbutton station, a limit switch, a pressure switch, a float-switch, or etc.? Or—to put it another way—when I'm specifying or ordering a magnetic motor starter, do I want it *with* or *without* a fuse block for providing control circuit protection? NEC Sec. 430-72(b) contains the answer to that question—if you can extract it from all the words, cross-referencing, and Table 430-72(b), with its further referencing. (It's a merry-go-round!)

A very detailed explanation of the rules of Sec. 430-72 is given on pages 883 to 891 of our *McGraw-Hill's National Electrical Code Handbook*. Certainly, a technical explanation that covers 9 pages in a textbook cannot be covered in detail in this brief article. But we are concerned here with warning electrical design and installation personnel that they can be (and have been) held responsible for satisfying mandatory NEC regulations that are obscure and convoluted and can be readily orchestrated by plaintiffs' attorneys to create the basis for convincing a jury that the cause of an accident was the ignorance and malpractice of the people who made the electrical design and installation.

A number of background considerations are as follows:

1. First, it must be carefully noted that the rules of Sec. 430-72(b) are about *"overcurrent* protection" for the circuit *conductors* and *not* for the operating coil of the starter.

2. These rules are concerned with "overcurrent protection" for coil-circuit conductors, which are carrying *less than* 2A in over 99 percent of all cases—even for starters up to NEMA 8 size and larger. The "load current" of any coil is the amount of current drawn.

3. Reference to NEC Table 430-72(b) invariably raises more questions than it answers. The wording of part (b) of Sec. 430-72 and the relation of the "BASIC RULE" [which is the first sentence of part (b)] and Exceptions No. 1 and No. 2 to Table 430-72(b) is cumbersome, to say the least. The Table reference to "control circuit conductors" that are "larger than No. 10" often puzzles people. Sec. 430-72(a) says that Sec. 430-72 covers only those control circuits that are "tapped" within a starter. How often are such circuits made with "larger than No. 10" wire?

4. Table 430-72(b) requires Nos. 18 and 16 copper control-circuit conductors to be protected at 7A and 10A, respectively, under Columns A and C. Sec. 725-16(a) permits No. 18 and No. 16 conductors to be used for motor control circuits and part (b) of that section lists the "fixture wires" that are *required* to be used for such circuits. Then NEC Table 402-6 shows that *all* No. 18 fixture wires have an ampacity of 6A and *all* No. 16 fixture wires have an ampacity of 8A. Why not 7A and 10A, *or* 6A and 8A in *both* Secs. 402-5 and 430-72? No big deal, especially when we're still talking about less than 2A in 99 percent of control circuits. But this is just another example of how the NEC suffers from lack of consistency.

On the basis of what we've said and the references we made, we do hope that all design and installation people will make an extraordinary effort to study these very important rules that critically impact success in the electrical field. As a single solution to the whole problem of control-circuit protection, we say: *Provide fuse protection for all motor starter (and contactor) coil circuits.* Or be fully prepared to defend any omission of such protection.

Proper Transformer Protection Requires "Coordination"

Secs. 110-10, 240-3, 240-100, and 450-3(b). **Satisfying NEC requirements for transformer overcurrent protection *without* nuisance tripping requires coordination of transformer inrush and damage points with the overcurrent protective device's time-current trip curve.**

An important element in the design of transformer layouts for electrical distribution systems is the provision of effective overcurrent protection for each transformer. With regard to transformer overcurrent protection, careful observance of the rules of NEC Sec. 450-3(b) and coordination of the protective device trip-curve with transformer inrush and damage points will ensure compliance with the Code and complete adequacy for transformers rated up to 600V on their primaries.

As given in Sec. 450-3(b), there are two acceptable methods for protecting transformers against overcurrent. They are:

1. *Primary protection only.* Where primary overcurrent device is considered to also protect the secondary and no secondary protection is required, and

2. *Primary and secondary protection.* Where the primary feeder overcurrent device only protects the transformer primary and another overcurrent device provides protection for the transformer secondary.

It is worth noting that additional requirements for conductor and panelboard protection covered by Secs. 240-3 and 384-16, respectively, must be satisfied. Compliance with 450-3(b) does not ensure that minimum requirements for conductor and panelboard overcurrent protection are met.

Code requirements

Primary protection only. The basic rule of Sec. 450-3(b)(1) as applied to protection of a dry-type transformer used for power stepdown to 208Y/120V, 3-phase, 4-wire is shown in Figure 1. Overcurrent protection for the dry-type transformer is provided by fuses or CBs rated at

Transformer Inrush Current

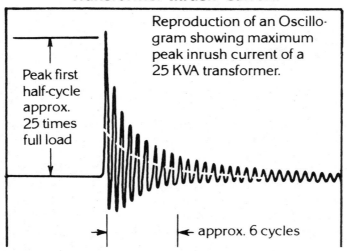

Peak first half-cycle approx. 25 times full load

Reproduction of an Oscillogram showing maximum peak inrush current of a 25 KVA transformer.

← approx. 6 cycles

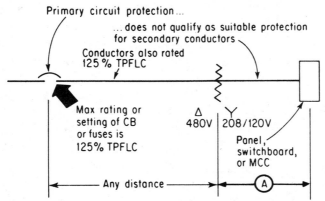

Primary circuit protection...

...does not qualify as suitable protection for secondary conductors

Conductors also rated 125 % TPFLC

Max rating or setting of CB or fuses is 125% TPFLC

Δ 480V | Y 208/120V

Panel, switchboard, or MCC

← Any distance → ← (A) →

Distance "A" from transformer to panel is limited to 10 or 25 ft, subject to the requirements of Exceptions 2 and 3 to Section 240-21. If overcurrent protection is placed at the transformer secondary connection to protect secondary conductors, the circuit can run any distance to the panel. If the panel fed by the transformer secondary is a lighting panel and requires main protection, the protection must be on the secondary [Section 384-16(d)].

Figure 1 The requirements for protecting a transformer rated 600V or less against overcurrent using primary-only protection are given in Sec. 450-3(b)(1). In that section the NEC recognizes an overcurrent protective device set or rated at not more than 125 percent of the TPFLC on the transformer's primary side as adequate protection for both the primary and secondary.

not more than 125 percent of the transformer primary full-load current (TPFLC).

As indicated, a CB or set of fuses rated not over 125 percent of the TPFLC provides all the overcurrent protection required by the NEC for the transformer. This overcurrent protection is in the feeder circuit to the transformer, and it is logically placed at the supply end of the feeder. When the correct maximum rating of transformer protection is selected and installed at the supply end of the feeder, the feeder conductors must be sized so that the CB or fuses selected will provide proper protection as required for the conductors. The ampacity of the feeder conductors must be at least equal to the ampere rating of the CB or fuses, unless Exception No. 4 to Sec. 240-3 is satisfied. That is, when the rating of the overcurrent protective device selected is not more than 125 percent of the TPFLC, the feeder conductors may be considered properly protected if the ampacity of the feeder conductor is at least equal to the TPFLC and is such that the selected protective device is "the next higher standard device rating" above the conductor ampacity. In addition, the secondary conductors must be protected as required by Article 240. Those concepts are intended to be communicated by the FPN following the first paragraph of Sec. 450-3, which refers to Sec. 240-3 and 240-100 for overcurrent protection requirements of the conductors.

The Exceptions to the basic rule of Sec. 450-3(b)(1) recognize higher maximum ratings for primary-only protection under certain conditions.

For transformers whose TPFLC is rated 9A or more, where 1.25 times TPFLC does not correspond to a standard rating of protective device, the next higher standard device rating from Sec. 240-6 is permitted. This only applies to nonadjustable type CBs and fuses. Adjustable-type CBs *must* be set at no more than 125 percent of the TPFLC where primary protection only is provided.

For transformers whose TPFLC is rated less than 9A, the rating of the primary protective device may be set or rated at *not more than* 167 percent of the TPFLC where primary protection only is provided. And for transformers whose TPFLC is less than 2A, the primary overcurrent protective device may be set or rated at *not more than* 300 percent of the TPFLC where primary protection only is provided (Figure 2).

Primary and secondary protection. Another way to protect a 600V transformer is described in Sec. 450-3(b)(2). In this method, the transformer primary may be fed from a circuit which has overcurrent protection (and circuit conductors) rated up to 250 percent (instead of 125 percent) of the TPFLC; but, in such cases, there must be a protective de-

**A transformer with rated primary
current of *9 amps or more* . . .**

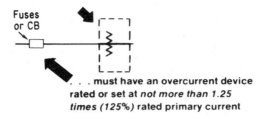

Fuses
or CB

. . . must have an overcurrent device
rated or set at *not more than 1.25
times (125%)* rated primary current

NOTE: Where 1.25 times primary current does not correspond to a standard rating of protective device. the next higher standard rating from Section 240-6 is permitted.

**A transformer with rated primary
current of *less than 9 amps* . . .**

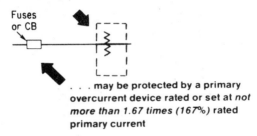

Fuses
or CB

. . . may be protected by a primary
overcurrent device rated or set at *not
more than 1.67 times (167%)* rated
primary current

**A transformer with rated primary
current of *less than 2 amps* . . .**

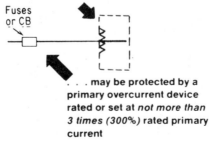

Fuses
or CB

. . . may be protected by a
primary overcurrent device
rated or set at *not more than
3 times (300%)* rated primary
current

Figure 2 For transformers whose TPFLC is greater than 9A, Exception No. 1 to Sec. 450-3(b)(1) permits use of "the next standard rating" of protective device when the calculated value of transformer overcurrent protection does not correspond to a standard protective device rating as covered in NEC Sec. 240-6. *But,* this applies *only* to fuses and nonadjustable-trip CBs. Where a CB with an adjustable long-time trip setting is used, the CB must not be adjusted above the 125 percent *maximum* established by the basic rule of Sec. 450-3(b)(1).

vice on the secondary of the transformer, to provide secondary protection, and this device must be rated or set at not more than 125 percent of the transformer secondary full-load current (TSFLC) as shown in Figure 3. This secondary protection must be located right at the transformer secondary terminals or not more than the length of a 10 or 25 ft. tap away from the transformer, and the rules on tap conductors (Sec. 240-21, Exception No. 2 or 3), as well as the rules on panelboard protection (Sec. 384-16), must be fully satisfied.

Under certain conditions, Sec. 450-3(b)(2) also recognizes coordinated thermal overload protection (integral protection provided by the transformer manufacturer), without a separate secondary overcurrent device, as providing the transformer protection required by the Code (Figure 4).

The Exception to Sec. 450-3(b)(2) modifies the basic 125 percent maximum rating for the transformer secondary overcurrent protection under certain conditions. Where 1.25 times the TSFLC does *not* correspond to a standard rating of protective device (as covered in Sec. 240-6), the next higher standard rated protective device may be used as secondary protection for transformers whose TSFLC is 9A or greater. *But,* this applies to nonadjustable CBs and fuses only. Adjustable-type CBs must be set *at not more* than 125 percent of the TSFLC. And, for transformers whose TSFLC is less than 9A, the maximum rating of the secondary overcurrent protective device may be set or rated at *not more than* 167 percent of the TSFLC.

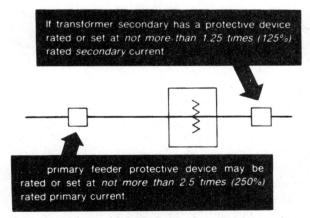

If transformer secondary has a protective device rated or set at *not more than 1.25 times (125%)* rated *secondary* current

primary feeder protective device may be rated or set at *not more than 2.5 times (250%)* rated primary current

Figure 3 The basic rule of Sec. 450-3(b)(2): primary and secondary protection. The FPN following the first paragraph of Sec. 450-3 is intended to serve as a reminder that compliance with the requirements for transformer overcurrent protection must be carefully coordinated with the requirements of Sec. 240-3 to assure that the primary and secondary *conductors* are protected as required by that section and located as covered in Sec. 240-21.

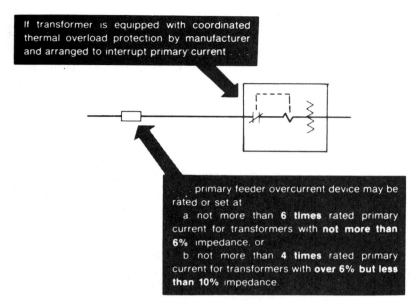

If transformer is equipped with coordinated thermal overload protection by manufacturer and arranged to interrupt primary current . . . .

. . . primary feeder overcurrent device may be rated or set at

a. not more than **6 times** rated primary current for transformers with **not more than 6%** impedance. or

b. not more than **4 times** rated primary current for transformers with **over 6% but less than 10%** impedance.

Figure 4   The second paragraph of Sec. 450-3(b)(2) recognizes an application with primary protection only where the transformer is manufacturer-equipped with an integral thermal overload protective device that will open the primary circuit as specified according to the transformer's rated impedance.

Coordination

In addition to determining the proper rating or setting for the transformer overcurrent protective device, coordination of the protective device time-current trip-curve with the transformer inrush and damage points is necessary to prevent nuisance tripping and assure that the transformer is adequately protected.

A 125-percent-rated primary protective device is generally considered adequate to accommodate transformer inrush if the device can carry a current value of 12 times TPFLC for 0.1 second without opening the circuit. Some transformer manufacturers, however, have indicated that in addition to coordination at 12 times TPFLC for 0.1 second, the protective device trip-curve should be evaluated at 25 times the TPFLC for 0.01 second and at 3 times for 10 seconds.

Because transformer inrush current may reach (and even exceed) a value of 25 times the TPFLC in the first half cycle as the core is magnetized, this inrush point should be considered when selecting the primary overcurrent protective device. If the primary overcurrent protective device is a fuse, the selected fuse should be such that the minimum melting-current at 0.01 sec. is greater than 25 times the TPFLC. If the primary device is a CB, the selected CB should be such that its magnetic unlatching-current is greater than 25 times TPFLC for 0.01 sec. Additionally, because loads supplied by the transformer may have current requirements of several times full-load current (e.g., motor inrush), the primary protective device should be capable of carrying 3 times the TPFLC for 10 sec. without opening the circuit.

Coordination of the protective device trip characteristics with the transformer damage points is necessary to ensure that the selected protective device will interrupt the circuit before the current can reach a value that will cause a significant and irreversible degradation of the transformer insulating material. Because such stresses will adversely affect the service life of the transformer, good design practice dictates careful consideration of transformer damage points, as well as coordination with the recommended inrush clearance points.

When thermal-magnetic circuit breakers are sized at 125 percent of rated TPFLC for transformer primary-only overcurrent protection, coordination with transformer inrush and damage points may be difficult to achieve.

For example, consider the application shown in Figure 1. Assuming that this transformer is rated at 112.5kVA, what would be the maximum rating of protective device required by the Code for primary-only protection? As covered in Sec. 450-3(b)(1), the maximum rating of the transformer primary protective device must be 1.25 times the TPFLC. For a 3-phase transformer, TPFLC is determined from the transformer kVA rating as follows:

$$\text{TPFLC} = \text{kVA} \times 1000/1.73 \times \text{Phase-to-Phase Voltage}$$

$$\text{TPFLC} = 112.5 \times 1000/1.73 \times 480$$

$$\text{TPFLC} = 112500/830$$

$$\text{TPFLC} = 135.5\text{A}$$

Per Sec. 450-3(b)(1), the maximum rating of the primary overcurrent protective device must be:

$$1.25 \times \text{TPFLC} = \text{Maximum Rating}$$

$$1.25 \times 135.5\text{A} = 169.375\text{A}$$

For nonadjustable CBs and fuses, Exception No. 1 to Sec. 450-3(b)(1) recognizes use of the next higher standard rating of protective device, as covered in Sec. 240-6, for transformers whose TPFLC is 9A or more where the "calculated" ampere rating for the protective device does not correspond to a standard rating of protective device. Therefore, the maximum rating of fuse or nonadjustable CB (adjustable-type CBs *must* be *set* at *no more than* 125 percent of the TPFLC) must be 175A (next standard size) to protect the 112.5kVA transformer where only primary protection is provided.

Figure 5 shows the trip-curve for a typical 175A molded-case CB with plots of the inrush clearance points and damage points for a 480VAC to 208Y/120VAC dry-type transformer.

Generally, the magnetic operation of a standard molded-case thermal-magnetic circuit breaker occurs at approximately 10 times (maximum) its trip rating. Although this curve reveals that this breaker would not open the circuit at 12 times TPFLC for 0.1 second,

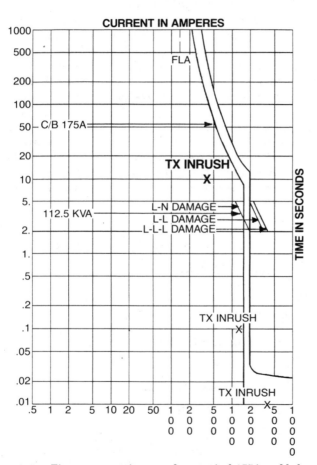

Figure 5 Time-current trip curve for a typical 175A molded-case circuit breaker with transformer inrush-clearance points plotted at 25x's TPFLC for 0.01 sec., 12x's for 0.1 sec., and 3x's for 10 seconds for a 112.5kVA, 3-phase, 480VAC to 208Y/120VAC transformer. Transformer damage points are also shown.

notice that the thermal magnetic circuit breaker's instantaneous pickup at 1,750A is well to the left of the first cycle inrush clearance point of 25 times TPFLC for 0.01 second (3,387A). We can see graphically that worst-case transformer inrush could cause nuisance trip of the circuit breaker.

To accommodate this worst-case inrush condition, the circuit breaker would have to be up-sized to 250 percent of TPFLC to allow for magnetization of the transformer without nuisance tripping of the circuit breaker. In addition, to comply with the other requirement of Sec. 450-3(b)(2), secondary protection rated or set at not more than 125 percent of the TSFLC must be provided. And the primary conductors would have to be up-sized to assure compliance with Sec. 240-3.

Figure 5 also plots three sets of time-current damage points for the 112.5 KVA 3-phase dry-type transformer with a 480VAC delta primary and a 120/208 grounded-wye secondary. The left-side set of damage points describe the combinations of time and current which, during a line-to-neutral fault, could cause damage to the 112.5 KVA transformer. Moving to the right, the next line represents the line-to-line damage points, and the final line, the 3-phase damage points. Note that the total clearing curve of the thermal-magnetic circuit breaker falls to the right of the line-to-neutral transformer damage points. We can see graphically that should a line-to-neutral fault of 1,200 to 1,749A occur, damage to the transformer may result before the circuit breaker clears.

Figure 6 plots the same 112.5 KVA transformer inrush and damage points as well as the characteristics of an RK-5, 175A, dual-element, time-delay fuse. Note that the minimum melt values of the 175A fuse fall well to the right of the transformer inrush points, and that the total clearing values give adequate coordination with line-to-line and line-to-line-to-line transformer damage points. However, the transformer could still be damaged by line-to-neutral faults.

Figure 7 shows the trip-characteristics of an RK-5, 150A, dual-element, time-delay fuse, sized smaller than the *maximum* allowed by 450-3(b). Notice that coordination with first-cycle inrush is maintained while coordination with all damage points is adequate.

It is worth noting that, although Sec. 450-3(b) establishes "maximum" values for transformer overcurrent protective devices, it is permissible to use a device that is rated *less than* the maximum, provided all other Code rules, particularly Sec. 220-10, are satisfied. That is, if the protective device that provides transformer primary protection is to supply a *continuous* load, then the protective device must be rated such that the continuous load would not be greater than 80 percent of the breaker rating (or the breaker must be rated at 125 percent of the continuous current). In this example, if all load supplied from this

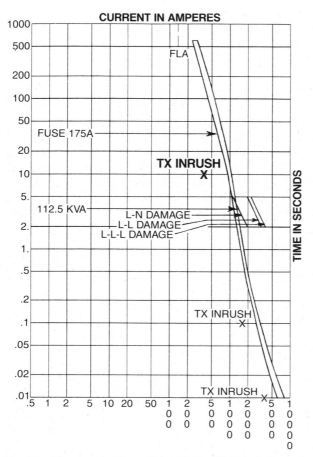

CURRENT IN AMPERES

TIME IN SECONDS

FLA

FUSE 175A

TX INRUSH
X

112.5 KVA

L-N DAMAGE
L-L DAMAGE
L-L-L DAMAGE

TX INRUSH
X

TX INRUSH
X

Figure 6 Time-current curve for a 175A, RK-5 dual-element time delay fuse with the 112.5kVA transformer's inrush-clearance and damage points plotted.

transformer were general lighting in an office building or industrial facility, which is considered to be a "continuous load," the rating of the protective device must not be less than 125 percent times the TPFLC or 169.375A. If the total noncontinuous primary load *plus* 1.25 times the continuous primary load is 150A or less, then use of the 150A fuse would satisfy the rule of Sec. 220-10(b).

The NEC contains provisions considered necessary for safety and per Sec. 90-1(c), it is not intended as a design guide. We can conclude from our calculations and observations that when sizing thermal magnetic circuit breakers for transformer primary-only overcurrent protection at the maximum rating permitted by Sec. 450-3(b) of the NEC, coordi-

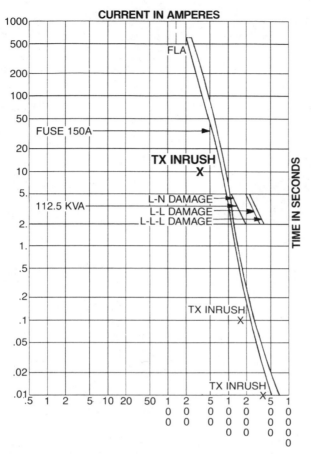

Figure 7 Time-current curve for a 150A, RK-5 dual-element time delay fuse with the 112.5kVA transformer's inrush clearance and damage points plotted.

nation of the circuit breaker with transformer first-cycle inrush may be difficult to achieve and nuisance trip upon transformer energization may occur. If a larger circuit breaker is used, transformer secondary overcurrent protection must be provided. In addition, good design practice may require sizing RK-5 dual-element time-delay fuses smaller than the NEC maximums in order to achieve adequate primary-only fused transformer protection.

Grounding of Submersible Pumps

Sec. 250-43(k). A discussion of practical methods for complying with NEC rules on grounding of submersible pumps and assuring pump service life.

The 1990 National Electrical Code (NEC) contains a new requirement for grounding of "motor-operated water pumps including the submersible type" [Sec. 250-43(k)]. Previously, there were many applications of submersible pumps that were not required to be grounded because they were "isolated or guarded." But there were many other installations where the submersible pumps *should* have been grounded—such as those subject to potential human contact, that is, where installed in ponds, lakes, cisterns, sumps, etc.—but they were not. This rule now clearly and definitely requires grounding of *all* submersible pumps.

The importance of this new rule is not lost on those who have to maintain, troubleshoot, and repair submersible pumps. There are many situations where pump maintenance personnel could be exposed to a shock or electrocution hazard if a faulted pump motor is not grounded.

For example, there are many instances where a pump is removed from the well and laid on an insulator (usually a piece of plywood). Although not recommended as good practice, many times power will be applied to make a measurement of current, voltage, etc. If the ungrounded pump motor is faulted to the case, and the individual touches the case, that person would receive a shock or be electrocuted (Figure 1). While such an incident can be avoided by always checking the case for voltage to ground before touching the pump, a low-impedance return path provided by an equipment grounding conductor would allow sufficient current flow to trip the overcurrent protective device.

Where large pumps are used—for example, 230VAC, 3-phase, 40hp—proper rotation of the pump shaft is normally verified prior to lowering the pump into the hole. If the shaft rotation is reversed, the reverse torque created by these "giants" can unscrew the drop pipe in seconds, which generally results in a very expensive pump and motor dropping to the bottom of the well. To assure proper shaft rotation, such pumps are usually "bumped" (jogged) to determine proper shaft rotation. During this procedure, if there is a faulted conductor or motor winding in the pump assembly, the personnel handling the pump would also be exposed to a shock or electrocution hazard.

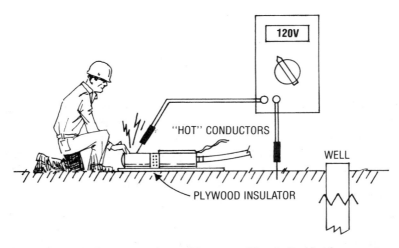

Figure 1   Troubleshooting a submersible pump with a faulted hot leg or motor winding can be lethal. Use of a properly sized equipment grounding conductor will help to assure automatic clearing of a fault and provide a greater level of safety for anyone maintaining or repairing the pump.

There are other instances where a maintenance person would be put at risk if an ungrounded submersible pump were faulted. As shown in Figure 2a and 2b, testing of a pump in an insulated pit—either above or below ground—would expose the tester to a potential shock or electrocution hazard. All pump testing stations should be fitted, by a certified (licensed) electrician, with a full-sized equipment grounding conductor which should be attached to the pump before it is lowered into the water. That way, if the motor or a "hot" conductor is faulted, there should be adequate current flow to operate the circuit protective device, thereby removing the dangerous voltage from the pump case.

Although most wells are large uninsulated concrete tanks, even those tanks that are in continuous contact with moist earth present a potential hazard. As we all know, current flow through the earth will virtually never be sufficient to cause operation of a circuit's overcurrent protective device. However, the earth will generally conduct enough current to electrocute a human being (it only takes about 100mA to kill a human; it takes more than 15 or 20A to trip a branch circuit protective device). With that in mind, it becomes apparent that grounding of *all* submersible pumps is the only certain method that will eliminate potential shock and electrocution hazards where a pump becomes faulted.

Another problem associated with ungrounded submersible pumps is damage due to lightning strikes. One might think that an *un*grounded submersible pump should be less likely to be damaged by lightning.

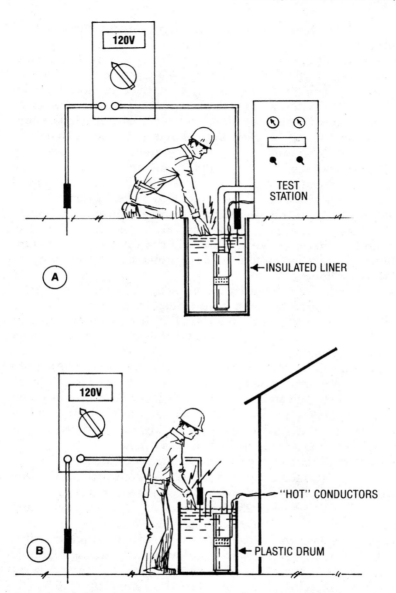

Figure 2   Whether the pump is in an insulated pit (A, above) or in an insulated tank (B), the maintenance person would risk electrocution or shock if a fault occurred and the pump were not grounded.

But so-called lightning-induced current surges destroy thousands of ungrounded submersible pumps every year.

Where a strike occurs on the neutral conductor, the resulting voltage drop will be directly proportional to the resistance of the neutral conductor and the grounding electrode system; the current induced by the difference of potential across that resistance will be proportional to the voltage amplitude and the ohmic value of resistance in the neutral conductor/grounding electrode system and connections. By lowering the resistance to ground, it is possible to reduce the voltage drop and the amplitude of the lightning-induced current. Connection of the pump to the equipment ground bus will place the submersible pump— which is deep underground and surrounded by water in contact with earth—in parallel with the grounding electrode connection to earth. That will reduce the overall ground resistance of the grounding electrode system and therefore the amplitude of the voltage spike and induced current, which will afford greater protection to the pump as well as other appliances.

Where a lightning strike enters the building on a "hot" conductor, high-quality surge arresters can be provided to greatly reduce the likelihood of damage due to such a strike. But, it is worth noting that ground resistance of the grounding electrode system will also affect performance of the surge arresters.

Although connection to the grounding electrode system through a properly sized equipment grounding conductor can serve to minimize damage due to lightning strikes, there is still some concern for pump damage. This relates to present field practice—especially in residential and rural installations—where only a single ground rod is provided and no test of ground resistance is performed.

There is no doubt that such practice is contrary to the NEC. But, inspectors very rarely enforce the rule of Sec. 250-84, which specifically requires more than one ground rod, unless a ground resistance measurement indicates a value of 25 ohms or less. Because it is virtually impossible to achieve a ground resistance value of 25 ohms or less in many areas of the eastern United States when using a single rod, it is safe to say that nearly every such installation with a single ground rod—either supplementing an underground piping system or serving as a building's grounding electrode—is in violation of the NEC.

As indicated, in applications where a single rod is used, the significantly lower ground resistance of the submersible pump will be in parallel with the higher ground resistance of the rod electrode. Refer to Figure 3. Assuming that the single driven ground rod has a ground resistance measurement of approximately 40 ohms, when it is placed in parallel with the pump and well casing which generally has a ground resistance of 5 ohms, the lightning-induced current would di-

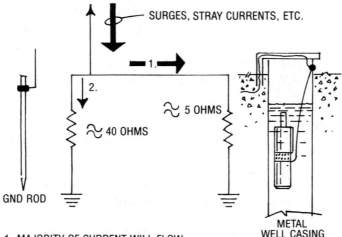

1. MAJORITY OF CURRENT WILL FLOW
   TO WELL CASING ON EQUIPMENT GROUNDING CONDUCTOR.
2. ONLY A SMALL FRACTION OF TOTAL CURRENT WILL FLOW
   TO GROUND ROD.

Figure 3 Division of lightning-induced surge currents between a single rod electrode and a pump/well casing. Assuring that the ground resistance of the building's grounding electrode system is as close as possible to that of the pump/well casing (typically, about 5 ohms) will serve to provide a more "even" split of the current to ground. The connection to the metal well casing is only specifically required for "Agricultural Buildings" [Sec. 547-8(d)], but should be provided for *all* metal well casings.

vide itself between the pump and the rod at an 8 to 1 ratio. Although it is agreed that the submersible pump should be grounded, it is not desirable for the pump to carry so much of the lightning-induced current because that current will tend to overstress the grounding conductor and connections. To minimize the amount of current carried by the pump's equipment grounding conductor and connections, the ground resistance value of the grounding electrode system must be reduced so as to match, as closely as possible, the typical 5 ohm ground resistance of the pump. That will permit a more equal division of the surge currents.

Selecting and connecting the equipment grounding conductor

The equipment grounding conductor now required for all submersible pumps must be sized in accordance with the rules of NEC 250-95 and Table 250-95. To the pump installer, this means the equipment grounding conductor must be a minimum No. 14 AWG copper conductor for a 15A circuit; NO. 12 AWG copper for a 20A circuit; and No. 10

AWG copper for all other circuit ratings up to 60A. Although those size conductors are adequate to assure sufficient current flow during a fault condition, given the fact that this grounding conductor will generally carry a fair amount of current should a lightning strike occur, I feel those conductor sizes are too small. I recommend a No. 6 copper as the minimum size conductor for grounding of a submersible pump. It is important to note that this is my opinion. The NEC would accept an equipment grounding conductor sized according to the rule of Sec. 250-95 and Table 250-95. Use of a larger conductor is completely voluntary.

The equipment grounding conductor may be bare, covered, or insulated. If covered or insulated, grounding conductors in sizes up to and including a No. 6 must be identified by an overall green coloring or green coloring with one or more yellow stripes.

That was the easy part. Connection of the grounding conductor presents another challenge.

Today, there are very few submersible pump motor manufacturers that have incorporated an equipment grounding conductor in the motor-lead assembly. And it may be another 1 or 2 years before those pump motors are brought to market. In the meantime, the required grounding conductor will have to be field-attached to the motor frame. Although not the most desirable situation, if good practices are followed and continuity of the ground return path is verified, such an installation should be capable of providing the required low-impedance return path.

Present designs of submersible pumps are such that the motor-lead guard (running down the side of the pump end) for those pumps with 2-wire motor leads should have adequate room to accommodate the grounding conductor (see Figure 4). Installation of the grounding conductor within the motor-lead guard will generally not be possible where the pump has a 3-wire motor. And some larger pumps, such as the 40hp unit previously mentioned, do not have motor-lead guards. In either of those cases, the grounding conductor will have to be secured (using tape or plastic ties) in close proximity to the hot conductors to prevent damage during lowering or raising of the pump and minimize return path impedance.

The point of connection for the equipment grounding conductor is also shown in Figure 4. I use one of the motor studs. However, before using a specific stud, I verify continuity between the motor case and the stud using an ohmmeter set on R × 1. A reading of less than 1 ohm is what should be measured. If a higher value is recorded, do not use that stud. If continuity cannot be found between any stud and the motor case, then it is necessary to contact the manufacturer for further instructions (Figure 5).

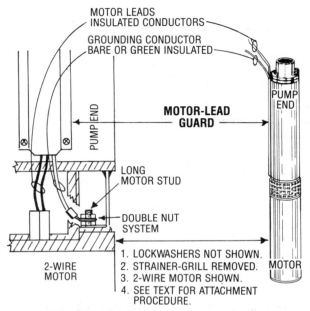

MOTOR LEADS
INSULATED CONDUCTORS

GROUNDING CONDUCTOR
BARE OR GREEN INSULATED

PUMP END

MOTOR-LEAD GUARD

PUMP END

LONG MOTOR STUD

DOUBLE NUT SYSTEM

2-WIRE MOTOR

1. LOCKWASHERS NOT SHOWN.
2. STRAINER-GRILL REMOVED.
3. 2-WIRE MOTOR SHOWN.
4. SEE TEXT FOR ATTACHMENT PROCEDURE.

MOTOR

Figure 4 Details for field connection of an equipment grounding conductor to a submersible pump.

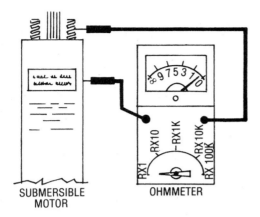

SUBMERSIBLE   OHMMETER
   MOTOR

STUD-TO-FRAME CONTINUITY TEST
MUST INDICATE LESS THAN ONE OHM

Figure 5 Always verify continuity between the motor case and the stud selected for connection of the equipment grounding conductor.

Once the point of connection is determined, the method of connection must be considered. As covered by the NEC, if the conductor to be terminated is larger than No. 10, it must first be joined to some type of lug- or barrel-type pressure connector. Even though the NEC would permit a conductor No. 10, or smaller, to be wrapped around the stud and secured by a nut, I prefer to use a crimped connector. In actual practice, this provides additional protection against the grounding conductor pulling out from under the nut, which can cause misalignment of the motor shaft and possibly "bind" the pump.

Where a connector is provided at the point of attachment, consideration must be given to the connector material. Although stainless steel would seem to be best suited to such application, I have not yet found a manufacturer of stainless steel connectors. Commercially available tin/lead-plated copper connectors would probably be acceptable to the NEC, but may not last long in well water with a pH less than 7. Generally, I use a solid nickel connector. Those connectors are not usually a stock item and will have to be special-ordered. However, they offer maximum corrosion resistance to acidic water.

Where the stud is long enough, a double-nut arrangement such as that shown in Figure 4 should be used. A stainless steel nut is used to secure the grounding conductor connector to the existing nut. Two serrated-edge lock washers are used to prevent motor vibration from loosening the connection. For those cases where the stud is too short to use a double-nut connection, the conductor can be connected using a single nut and lock washer (see Figure 6).

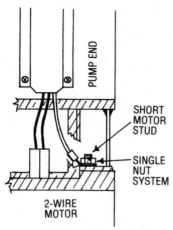

Figure 6 Where the stud is not long enough to allow for use of two nuts and washers, a single nut and washer can be used to secure the equipment grounding conductor to the motor as shown above.

The ground wire at the point of attachment must be kept within the profile of the pump to assure proper replacement of the strainer grille, which will generally have to be removed from the pump to get to the studs.

One final concern deals with grounding of a metal well casing. Although metal well casings are only specifically required to be grounded in Sec. 547-8(d), which covers "Agricultural Buildings," the wording of Sec. 250-42(a) would also seem to apply. Therefore, the noncurrent-carrying metal parts of all metal well casings should be grounded. This may be accomplished by use of a bonding jumper from the motor's equipment grounding conductor. Or, another equipment grounding conductor connected to the well casing and the ground bus in the panel that supplies the pump should also be acceptable. Again, sizing of that conductor in accordance with Sec. 250-95 and Table 250-95 would be acceptable, but consideration should be given to use of a larger-sized conductor to accommodate lightning surges.

Redundant Ground Path at Health-Care Facilities

Secs. 250-51 and Sec. 517-13(a), Ex. No. 1. Jacketing material must be listed as an equipment ground-return path where Type MC or Type AC is used to provide a redundant ground-return path.

In the 1990 NEC, a change was made in Sec. 517-13(a), Exception No. 1, that was intended to more clearly communicate that where Types MC and AC are used to supply receptacles or fixed equipment in patient-care areas of a health-care facility, the metallic jacket of such cables must be capable of providing the redundant ground path required by the basic rule. That is, in addition to containing an insulated, copper equipment grounding conductor, the metallic jacket itself must be "an approved grounding means."

The reason for the change in this section of the 1990 NEC is that the wording used in the 1987 NEC did not clearly convey the "belt-and-suspender" approach that is mandated by the basic rule and was intended to also be required when the permission given in the Exception was exercised. This Exception (Exception No. 1 to Sec. 517-11 in the 1987 NEC) read:

> *Exception No. 1: Metal raceways are not required where Type MC cable, Type MI cable, or Type AC cable with an insulated grounding conductor is used.*

That wording would accept the use of Type MC, Type MI, or Type AC cable for supplying fixed equipment or receptacles in patient-care areas of health-care facilities as long as the metal jacketed cable also contained "an insulated grounding conductor." And many people used the interlocked metal jacketed Type MC cable because it was the least expensive of the three recognized methods. However, this was not what the CMP had intended.

The metallic jacketing on Type MI cable is a continuous copper sheath—essentially, a copper tube—that is evaluated and listed as a ground-return path. The jacketing on all Type AC cables (except Type ACL) is an interlocking spiral wound metal jacket with a No. 16 aluminum bonding strip that has been evaluated and listed as a ground-return path. However, while the jacketing on *certain* Type MC cables *is* listed as an equipment grounding means, the metallic jacketing on *all* Type MC cables is *not* so listed.

The UL "lists" three different designs of Type MC cable: (a) interlocked metal tape, (b) corrugated tube, and (c) smooth tube. The metallic jacketing on those Type MC cables that have the smooth- or corrugated-tube-type jacketing *is* listed as an acceptable ground-return path. But, the metal jacketing on Type MC with the interlocking spiral wound metal jacket is *not* listed for use as an equipment ground-return path.

This is because under fault conditions any current flow will tend to follow convolutions or spiral turns of the metal jacket. This is actually a choke coil and presents an unacceptably high impedance to the flow of fault current and will not facilitate operation of the circuit overcurrent protective device as required by part (3) of Sec. 250-51. This is not a concern with interlocked armor Type AC cable because the No. 16 aluminum bonding strip required in Type AC cable effectively "shorts out" the turns of the metal jacketing and has been tested and shown to permit adequate current flow under fault conditions.

The wording of this rule as it now appears in Sec. 517-13(a), Exception No. 1, specifically addresses the concern for a redundant ground-return path and permits the use of a metallic jacketed cable to supply fixed equipment and receptacles in patient-care areas of health-care facilities *only* "where the outer metal jacket is an approved grounding means of a listed cable assembly." Therefore, unless the metal jacket on a Type MC cable is also listed as an equipment grounding means, such a cable assembly may not be used for supplying receptacles and fixed equipment in patient-care areas of health-care facilities.

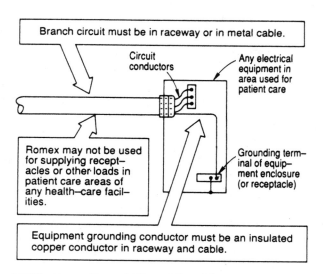

BUT, where Type MC or AC cables are used, the metallic cable jacket must be "approved" as a ground return path!

Another change in the wording of Sec. 517-13(a), Exception No. 1, was the deletion of the phrase "with an insulated grounding conductor." Although this phrase was dropped, it was not intended to indicate that an insulated equipment grounding conductor is no longer required. The only intent of this Exception is to recognize the use of the metallic jacket on Types MI, MC, and AC cable assemblies as the redundant ground path instead of the metallic raceway required by the basic rule of Sec. 517-13(a). *But,* such application of Types MC and AC cable is permitted only when the metallic cable jacket is "approved" (listed) for use as an equipment ground-return path. Therefore, the requirement for an insulated copper equipment grounding conductor given in the basic rule of Sec. 517-13(a) must still be satisfied.

Is "Bonding" Required for Hydromassage Bathtubs?

Secs. 680-4, 680-41, and 680-71. Definitions provide answers to questions on bonding required for spas, hot tubs, and hydromassage bathtubs.

Question: I have a few questions about the NEC-required grounding and bonding of a Jacuzzi installed indoors at a residence.

In a recent installation, I bonded the water piping and other noncurrent-carrying metal parts of the tub to the pump motor with a solid No. 8 AWG insulated copper conductor and grounded the motor with a No. 10 copper equipment grounding conductor (from Table 250-95 for a 40A protective device). My local electrical inspector would not accept this installation until I connected the No. 8 bonding conductor to a driven ground rod in the crawl space below the first-floor bathroom where the Jacuzzi was installed.

Afterwards, I discussed this with a few other contractors and they agreed that the NEC does *not* require attachment to a driven ground. But, they also said that the NEC doesn't require any of the bonding connections I made.

At this point I'm confused. What bonding does the Code require for such an installation and where is it covered?

Answer: The NEC addresses the installation of such equipment in Article 680. That article is divided into several subparts (A through G), which cover pools; spas and hot tubs; fountains; therapeutic-pools and -tubs; and hydromassage bathtubs. Of these various subparts, either Part D or Part G of Article 680 would apply to your installation.

In the 1984 NEC, spas, hot tubs, *and* hydromassage bathtubs were all covered by Part D of Article 680. And the rules of Sec. 680-41(d) and (e) in the 1984 NEC required the bonding of various noncurrent-carrying metallic parts with a solid No. 8 copper bonding conductor. Those requirements of the 1984 NEC have been revised slightly, but as they appear in the 1990 Code, the rules are still essentially the same. However, these rules now only apply to spas and hot tubs and *not* to hydromassage bathtubs.

One change in the 1987 edition of the NEC was removal of the term *hydromassage bathtub* from the heading of Part D and the establishment of a new subpart (Part G), which specifically covers hydromassage bathtubs as a separate issue. Although often overlooked, this change has eliminated all of the bonding requirements for equipment that meets the definition of "hydromassage bathtub" as given in Sec. 680-4. Therefore, determination of whether or not bonding is required must be based on the definitions provided at the very beginning of Article 680 in Sec. 680-4.

In Sec. 680-4, the Code gives definitions for a "spa or hot tub" and a "hydromassage bathtub." Although very similar in use, the Code-defined difference is based on whether or not the equipment in question is normally drained or not. If the "jacuzzi" in question is *not* normally drained after each use, it is a "spa" or "hot tub" and the requirements of Part D—including bonding of the designated noncurrent-carrying metal parts—must be satisfied. If the "jacuzzi" *is* normally drained after each use, it is a "hydromassage bathtub" and the two sections (Secs. 680-70 and -71) of Part G must be satisfied.

For a "hydromassage bathtub," Sec. 680-70 only requires that all circuits feeding electric equipment associated with the tub (e.g., pump, heater) be provided with personnel ground-fault protection. And, as stated in Sec. 680-71, all other electric equipment not directly associated with the tub is to be installed as would normally be required by the general rules given in Chapters 1 through 4. Compliance with those two rules is all that the Code requires for hydromassage bathtubs in the 1990 NEC.

As far as the connection to a driven ground rod goes, it is very difficult to understand where the inspector may have gotten that notion. It is equally difficult to comprehend what benefit is derived from such connection. The only real benefit of the Code-required connection to earth through a grounding electrode system is to provide for lightning discharge. The ground rod called for by this inspector is *not* required by any Code rule and really does nothing to enhance safety for the installation in question.

One last consideration is worth noting. Some hydromassage bathtubs are delivered with installation instructions that call for bonding as was previously required by the NEC for such units. Although the rules of Sec. 680-70 and -71 would *not* mandate such bonding, Sec. 110-3(b)—which requires listed equipment to be used in accordance with any instructions included in the listing or labeling—could be construed to require compliance with the manufacturer's installation instructions. To avoid any problems with an inspector and at the same time limit your legal exposure, always bond if the hydromassage bathtub manufacturer calls for bonding of specific noncurrent-carrying metal parts, even though the NEC does *not*.

Note 3 to Table 1, Chapter 9 of the NEC

"All of the same size" means the same square-inch area.

Question: The rule of Note 3 to Table 1 in Chapter 9 states:

Where the calculated number of conductors, ALL OF THE SAME SIZE, includes a decimal fraction, the next higher whole number shall be used where this decimal is 0.8 or larger.

My question is, "Does 'all of the same size' refer to AWG or area of square inches?" A No. 12 bare, a No. 12 TW, No. 12 THW, and No. 12 THHN are the same AWG size but are *not* the same overall size (in square-inch area). Additionally, dimensions given for conduit or tubing are given in square-inch area.

I had the following question on an electrical exam: If ten No. 8 RHW conductors with outer covering are installed in a 2-inch conduit, how many No. 8 XHHW conductors could you add to this raceway?

My answer:

Table 5—each No. 8 RHW (with cover) = 0.0854 sq. in.

Table 4—2 in. conduit (40% fill) = 1.340 sq. in.

Table 5—ten No. 8 RHW's (with cover) = 0.854 sq. in.

Subtracting one from the other gives remaining capacity.

Remaining capacity = 0.495 sq. in.

Table 5—a single No. 8 XHHW = 0.0456 sq. in.

.495 (remaining capacity) divided by 0.0456 = 10.85 conductors

Since the No. 8 RHW (with cover) and the No. 8 XHHW are not *all the same size*, the number of XHHW conductors should not be rounded up to the next whole number. Am I correct?

Answer: There are actually two questions here. The first deals with what is meant by the phrase "all of the same size" as used in Note 3 to Table 1 in Chapter 9 of the National Electrical Code (NEC). The second has to do with how this phrase relates to the electrical exam test question.

In regard to the meaning of the phrase "all of the same size," I would say that this refers to the overall cross-sectional area in square inches. As you correctly pointed out, conductors of the same AWG size can have a different overall size as measured in square-inch area depending on the type of insulation used. This is because with certain insulating materials less insulation is used due to their greater dielectric strength. And when such conductors are used, they physically occupy less space within the conduit, and therefore more conductors may be used within the conduit and still not exceed the 40 percent maximum allowable fill.

As far as the test question is concerned, although it is true that RHW (with outer covering) and XHHW would not be considered to be "all of the same size" in accordance with the explanation given above, all the XHHW conductors *would be* considered to be the "same size." That is, because the calculation involves determining the maximum number of XHHW conductors *only,* the rule of Note 3 to Table 1 in Chapter 9 *would* apply and the answer should be 11.

As indicated, the whole effort here is to determine the maximum allowable number of conductors that the Code would permit within the conduit. The rule of Note 3 to Table 1 in Chapter 9 is intended to address those situations where the calculated number of conductors is not a whole number. For the purpose of application and enforcement, this rule establishes the point at which the next whole number may be used when determining the maximum number of conductors that are "all the same size."

From the standpoint of the installing electrician, contractor, or engineer, this rule actually represents *permission* to go to the next whole number, even though the wording of the rule is mandatory in nature. That is, the rule would permit, but not require, an additional conductor within the conduit whenever the calculated number of conductors yields a decimal value of 0.8 or larger. And from the inspector's point of view, because the wording *is* mandatory, the maximum number of XHHW conductors that would be permitted by the Code within the 2-in. conduit must be taken to be 11.

Index